Statistical thinking for managers

Statistical thinking for managers

J. A. John
D. Whitaker
D. G. Johnson

CHAPMAN & HALL/CRC

Boca Raton   London   New York   Washington, D.C.

Library of Congress Cataloging-in-Publication Data

John, J. A.
   Statistical thinking for manager / J.A. John, David Whitaker, David G. Johnson.
      p. cm.
   Includes bibliographical references and index.
   ISBN 1-58488-248-4 (alk. paper)
   1. Mathematical statistics. I. Whitaker, David. II. Johnson, David G., 1943- III. Title.

QA276.12 .J375 2001
001.4'22—dc21   2001028747

Visit the CRC Press Web site at www.crcpress.com

Contents

Preface

Many managers, whether in manufacturing or processing industries, in service organisations, government, education or health, think that statistics is not relevant to their jobs. Perhaps it is because they see statistics as a collection of complicated techniques for which they see little application.

However, all business activities are subject to variability, and hence uncertainty, and managers have to make decisions in this environment. Managers need to understand the nature of variability. Otherwise they end up making wrong or inappropriate decisions, which can be very costly to their organisations. As well as understanding variability, good decision-making involves the use of meaningful information. Without good information, or data, managers have to resort to gut feelings or hunches, neither of which can be relied on. Statistics tells us how to deal with variability, and how to collect and use data so that we can make informed decisions. All managers need to be able to *think statistically* and to appreciate basic *statistical ideas*.

The aim of the book is to introduce you to this way of thinking, and to some of these ideas. Statistical thinking should be an integral, and important, part of any manager's knowledge. H. G. Wells* once said that statistical thinking would one day be as necessary for efficient citizenship as the ability to read and write. It should not be regarded as some separate activity that is done rarely, or left to others to do. Most managers would accept that they need a variety of different skills to inform and enhance their decision-making. For example, being able to correctly assess the financial implications of a certain decision and to see how it might affect staff and colleagues are undeniably important. But these are, in a sense, secondary issues. If you have failed to find the right course of action in the first place by not being able to fully appreciate the significance of the available data, then knowing what this action will cost to implement is of little consequence.

Unlike many books concerned with statistics for business or managers, this is not a book that emphasises mathematics or computation. Statisticians use mathematics

---

* *Mankind in the Making*, Chapman and Hall, 1914, page 204. This is not exactly what Wells wrote, but it is commonly paraphrased in this form.

extensively in their work, and the computer has become an indispensable tool for the analysis of data. However, our purpose is not to set out the mathematical basis of statistics, or to give a series of statistical techniques that require considerable computation. Statistical ideas can be understood with the minimum of arithmetical work, and this is what we aim to do in the book. This does not necessarily make the book simple or basic. Recognising how statistics, and statistical thinking, fits into the overall business picture is important, and requires careful thought and understanding. For example, being able to calculate a standard deviation, or the equation of a line of best fit, is not an end in itself. It is just a tool for describing data and is the beginning of a process of understanding and diagnosis, leading to effective decision-making.

A thorough statistical analysis can involve complex ideas and extensive computation for which specialised knowledge and computer programs are indispensable. At times, therefore, additional assistance from a qualified statistician will be needed; but this need has to be recognised. Knowing what you can confidently do for yourself, and where expert advice is needed, is one of the key characteristics of effective management.

The book has been set out as a workbook. You will see that on many pages there are questions for you to answer, examples for you to complete, and exercises for you to do. We strongly advise you to write your answers, thoughts, and solutions in the appropriate places in the book. This will help you keep all the material together, and allow you to revisit them at a later time. Supplementary information can be found on the book's website at http://www.crcpress.com.

It will be necessary to draw graphs, construct tables, and do some calculations. However, all these things can be done quite easily on the computer. In the book we make extensive use of the spreadsheet package Microsoft Excel. Spreadsheets are now standard on all computers and are, therefore, readily accessible. They are also widely used in business, and not just for statistical work. If you have not used a spreadsheet package before, or are new to Excel, you may find the introduction given in Appendix A a useful starting point. In the book you will be shown how to use Excel for analysing and plotting data. It will be well worthwhile working through the examples and exercises involving Excel on a computer. Excel is easy to use, and the more familiar you become with it the easier it will be to work through those sections involving its use.

One of the innovative features of this book is the inclusion of a number of "hands-on" exercises and experiments. It is another way that we hope to encourage a process of "learning by involvement." The questions that are contained within each chapter are there to actively involve you in the learning process by making you think about important issues to check your understanding. The experiments are a part of this same process, providing (we hope!) an active demonstration of a key concept through participation in a meaningful simulation. The most important of these experiments is the *beads experiment*, which is introduced in Chapter 1 and forms the basis of many of the important ideas introduced in the book. To get the most out of this experiment, it should be run in the classroom, so that you can see actual results occurring. If this is not possible we have provided a spreadsheet version of the experiment that you can run in Excel. A copy of this spreadsheet (*Beads.xls*) can be downloaded from the book's website. Instructions for using the spreadsheet are provided on the opening sheet. If you cannot participate in a classroom experiment, then substitute your own results from the beads spreadsheet.

In later chapters, you will find four other experiments that can either be run in the classroom or, in most cases, on your own. For two of these, the *dice experiment* and the *quincunx experiment*, there are spreadsheet alternatives that you can download and run so that you can simulate the actual experiment for yourself. Again, you can use your own results instead of those produced in the classroom or contained in the book. Each of these experiments demonstrates a key learning point. Think carefully about what these are, and try to relate them to real situations that you have experienced.

Most of the examples used in the book are based on our experiences working with a range of organisations throughout New Zealand and the U.K., or on those of our colleagues and associates. For confidentiality reasons we do not name those organisations, but have tried to preserve the realism of the examples. Although the context often relates to New Zealand, most, if not all, examples can be put into the context of any other country. We encourage you to think about similar problems or examples in your country.

The material in the book can be used in a number of different ways. All the material can be adequately covered in a 1-semester course of about 36 to 40 hours. Additional tutorial time should also be allowed, in which a selection from the exercises at the end of each chapter can be used. We do not see the need for separate computer laboratory sessions, as our aim throughout is to get an understanding of the role statistics plays in business, rather than with the arithmetical detail. The book can be used for a shorter course by skipping through some of the material previously covered elsewhere, or by omitting certain topics. We have used it on a short course by assuming prior knowledge of most of the material in Chapter 3, and by omitting Chapters 2, 9, and 13. Alternatively, or in addition, Chapter 10 could be omitted. For example, omitting Chapters 2, 9, 10, and 13 leaves sufficient material for a 1-semester course of about 22 to 25 hours, supplemented by 5 to 10 tutorials or problem classes. This is not to say that these chapters are not important, but simply that time constraints sometimes mean that not everything can be covered. It is better to work carefully through parts of the book, rather than attempt to cover it all superficially.

Many people have been involved, either directly or indirectly, in the development of this book. We particularly wish to acknowledge the contribution of Tim Ball, a consultant statistician in Wellington, New Zealand. Tim has freely made available material to us, and has been a consistent advocate of the importance of statistical thinking in business. We also wish to acknowledge Jocelyn Dale, Maurice Fletcher, Jock MacKay, Bruce Miller, Peter Mullins, and Peter Scholtes, all of whom have made contributions in various ways. We would also like to thank Kirsty Stroud, our editor at CRC Press, for her encouragement and suggestions on this book, and Karen Devoy for her excellent secretarial support and assistance. Over the years many undergraduate and MBA students, as well as managers and workers in industry, have been involved in courses based on all or part of this book. Their feedback and suggestions have been constructive and most welcome. Of course, any mistakes and errors are our responsibility alone.

Chapter 1

Variation

1.1 Variation

We live in a variable world. Managers have to make decisions in an environment subject to variability. Without understanding the nature of variability, managers face the danger of making wrong decisions and mistakes that can be very costly to their organisations. Variation exists in all processes, and reducing variation is the key to improving productivity, quality, and profitability. However, as Dr. W. Edwards Deming said, "the central problem in management and in leadership . . . is failure to understand the information in variation."[1] Statistics tells us how to deal with variation, and how to collect and use data so that we can make informed decisions.

In this chapter we shall discuss four problems involving variability. In each case, we would like you to think carefully what you would do if you were the manager involved, and to compare what you think with what the manager actually did. Then we shall run Dr. Deming's red beads experiment, and revisit these examples in light of this experiment. The lessons learned should give you an insight into the nature of variability, and an indication of what should be done.

1.2 Airport Immigration

The data in Figure 1.1 represent the number of passengers processed by immigration officers at an International Airport over the same fixed period of time. The passengers processed were those requiring a visa to enter the country.

Management expected their officers to process 10 passengers during this period.

The manager of immigration services, in reviewing these figures, was concerned about the performance of Colin, and was thinking about how best to reward Frank.

---

[1] *Out of the Crisis*, p. 309. Cambridge: MIT Center for Advanced Engineering Study, 1986.

Immigration officer	Number of passengers processed
Alan	9
Barbara	10
Colin	4
Dave	8
Enid	6
Frank	14

Figure 1.1   Customers Processed

Q1. What should the manager do about Colin?
Q2. How should she best reward Frank?

1.3   Debt Recovery

In Albion Cleaning, an office cleaning company, the level of unrecovered debt has been a cause of concern for some time. The data in Figure 1.2 give the percentage of invoices that have been paid within the due month. The company is aiming for a target recovery level of 80%. When the amount of recovered debt is much lower than this (in the low to mid 70%'s, for instance) the general manager visits all the district offices in New Zealand to remind managers of the importance of customers paying on time.

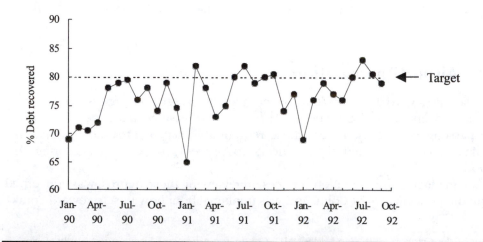

Figure 1.2   Percentage of Debts Collected Within Due Month

Q3. *Do you think the general manager has taken appropriate action?*

Q4. *Has the overall level of debt improved as a result?*

Q5. *What would you do, and why?*

1.4 Fire Insurance

The managing director of a company received a letter from its insurance company stating that unless there was a drastic reduction in the frequency of fires in the company's premises, the insurance company would cancel the policy. (Source: Deming, *Out of the Crisis*.)

Q6. *What would you do?*

Q7. *What do you think the managing director did about this problem?*

Q8. *Do you think the insurance company is being wise in threatening to cancel the policy?*

1.5 Budget Deviations

The graph in Figure 1.3 gives the budget deviations for a particular company over a 16-month period. Budget deviations measure the difference between the amount budgeted and the actual amount, expressed as a percentage of the budgeted amount. The aim, of course, is to have a zero deviation, but some variation is inevitable. Most of the variation lies between −3% and 4%.

Q9. *How should the financial director react to the 7% deviation in March 1999?*

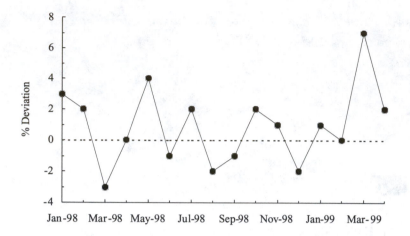

Figure 1.3   Budget Deviations

1.6   Red Beads Experiment

Dr. W. Edwards Deming was a prominent guru of quality management. In 1950 he started to teach the Japanese how to improve quality through the use of statistical quality control methods. His management philosophy had a big impact on industry in the United States and elsewhere from 1980 onwards. Mary Walton, in her book *The Deming Management Method*, gives a fascinating account of his methods, including his 14 points of management, and devotes a whole chapter to an important experiment on variability that Dr. Deming used to carry out in his teaching: the Red Beads Experiment.

We strongly recommend that you experience the beads experiment in a group situation. However, as an alternative, there is an Excel spreadsheet (*Beads.xls*) that you can use to simulate the results from a typical beads experiment. A copy of the spreadsheet is shown in Figure 1.10.

⬧ *Running the Experiment*

The tools for the experiment are

- a large number of small plastic beads, with the majority white and the others red,
- a paddle with 50 bead size holes in 5 rows of 10 each, as shown in Figure 1.10, and
- two plastic boxes, at least one of which is large enough for dipping the paddle.

It is important that there should be no departure from the following procedure for carrying out the experiment.

The beads are first placed in the larger plastic box. They now have to be thoroughly mixed by carefully tipping them into the other box, and then tipping them back into the first box. In doing this, each box should be held at the same constant height, say 8 cm, above the other box. A worker then takes the paddle and dips it into the box of beads, being careful to move the paddle along the bottom of the box so that the beads cover all the holes in the paddle. The paddle is removed slowly from the box so that no beads spill out of the paddle. There will now be 50 beads in the paddle, some of which will be white and some red. The number of red beads is counted and recorded. The red beads are defects. The purpose is to produce white beads, not red ones.

> **Q10.** *Why do you think it is important that the same procedure is followed each time?*

As the experiment is run, enter the results in the check sheet in Figure 1.4.

If you are not experiencing the experiment for yourself, run the beads spreadsheet to get a feel for what the results of the experiment will look like. The second sheet in *Beads.xls* (*Select beads*) allows you to simulate one run of the experiment by generating typical results for five workers over a 5-day period. At the same time, think carefully about what is involved in producing the beads in the paddle. Go through the process in your mind, and try to imagine what is likely to be happening. Finally, enter the data that you generate from *Beads.xls* into the table in Figure 1.4. Alternatively, simply copy the data shown in Figure 1.10.

> **Q11.** *What do you notice about the results of the experiment?*
>
> **Q12.** *What are the smallest and largest number of defects produced during the week? Explain why there is so much variability in the results.*

Worker	Day 1	Day 2	Day 3	Day 4	Day 5	Week Total
Total						

Figure 1.4   Pewerwight Bead Manufacturing Company

Q13. *If an incentive was offered to the best worker on any day, do you think it would help to motivate the workers? Would it be fair to give one of the workers a bonus?*

Q14. *If a worker was fired during the experiment, do you think this is fair? Why, or why not?*

Q15. *In this experiment do you think workers react better to positive or negative incentives?*

Q16. *What is the fundamental cause of defects in this experiment?*

Q17. *Whose responsibility is it to address this fundamental cause? Is it the workers, the supervisor, the inspectors, management, or who?*

Q18. *What would you do to reduce the number of defects in this experiment?*

✄ *Plotting the Data*

Trends, patterns, changes, and variability in the data are more easily seen in a plot, rather than a table, of the data. Plot the beads data in Figure 1.4 using the chart given in Figure 1.5. If you have used the beads spreadsheet to generate data, a graph of your results is contained in the *Select beads* sheet.

Calculate the average number of red beads, and draw a line on the graph to represent this average.

Q19. *Suppose we ran the experiment again. What results would you expect to get? What would the plot of the results look like?*

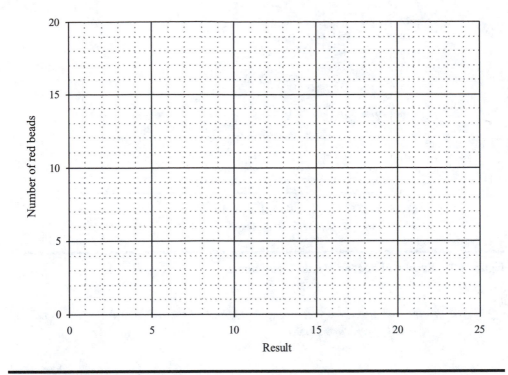

Figure 1.5   Plot of Beads Data

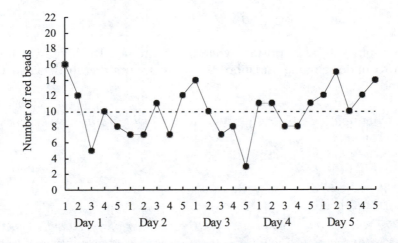

Figure 1.6   Results of First Beads Experiment

Figures 1.6 and 1.7 show the results of two other experiments carried out with different groups of students, using 1600 beads, of which 1200 are white and 400 are red. The overall average numbers of red beads are also drawn on the plots.

Q20. What can you say about the results of these two experiments?

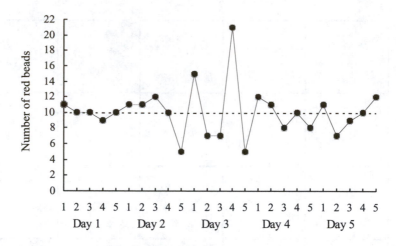

Figure 1.7   Results of Second Beads Experiment

> **Q21.** *In Figure 1.7, what do you think happened at the 4th result on day 3 when 21 red beads were produced?*

Some results will be surprising, while others are not. Deciding which is which is crucial to understanding variation. We shall discuss this in detail in Chapters 11, 12, and 13.

> **Q22.** *Having now seen the results of the experiment, would you be surprised if someone got*
> a. 50 red beads in the paddle?   Yes/No
> b. 40 red beads?   Yes/No
> c. 30 red beads?   Yes/No
> d. 0 red beads?   Yes/No
>
> **Q23.** *If we ran the experiment a very large number of times, with 1200 white and 400 red beads, what do you think the average number of red beads would be, and why?*
>
> **Q24.** *Does the evidence given in Figures 1.6 and 1.7 support your answer?*

☙ *Beads History*

A total of 425 results obtained from running the experiment 17 times are plotted in Figure 1.8, using 1600 beads, of which 1200 are white and 400 are red. The average values from each of these experiments are plotted in Figure 1.9. You can also use the sheet called *Repeat* within the beads spreadsheet to generate the results of a series of beads experiments and create a graph similar to that in Figure 1.9.

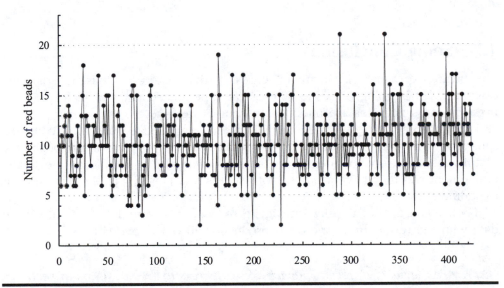

Figure 1.8   Results from 17 Beads Experiments

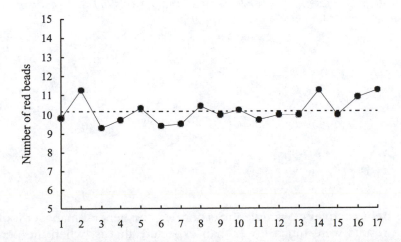

Figure 1.9   Averages from 17 Beads Experiments

> **Q25.** *Reconsider your answer to question 23. What, now, do you think the average number of red beads should be?*
>
> **Q26.** *Is your answer consistent with the results given in Figures 1.8 and 1.9? If not, why not?*

1.7   Some Conclusions

The beads experiment is, in many senses, a trivial exercise. It is transparently obvious what is going on, especially if you experienced it yourself. In practice, most business activities are much more complicated, yet there are many useful lessons that come from this experiment which can be applied to other, more complex, situations. Consider, for instance, the four examples we looked at in the beginning of this chapter. None of them is as straightforward or as simple as the beads experiment, yet they have a number of features that closely parallel those in our experiment.

Look again at the **airport immigration** data in Figure 1.1. Compare these data with the results from one day's production in our experiment.

> **Q27.** *Are they very different? Revisit your answers to the questions in Section 1.2 in light of the beads experiment. What conclusions do you now make?*
>
> **Q28.** *Suppose four more sets of data are available on the numbers of passengers processed by the six immigration officers. If Colin still processes the least number of passengers over these five time periods, what conclusions would you draw? What might you do next?*

In the **debt recovery** example in Section 1.3, the general manager took action when the results were much lower than expected; that is, when the percentage debt recovered was below about 75%.

Q29. *Was this sort of reaction apparent in the beads experiment? Can you see similarities between the two situations? Again review your answers to the questions in Section 1.3.*

In the **fire insurance** example in Section 1.4 the insurance company was threatening to cancel the policy.

Q30. *What can we say about the frequency of fires? What should the managing director do before anything else?*

Finally, in the **budget deviations** example in Section 1.5, the 7% deviation in the 15th month might seem to be a surprising result, but is it?

Q31. *Were the 21 red beads shown in Figure 1.7 surprising or unusual? How do we know? What do you think?*

1.8   Data for the Beads Experiment

The layout of the main sheet (*Select beads*) of the beads spreadsheet (*Beads.xls*) is shown in Figure 1.10. The display shows the contents of the paddle for Floyd on day 5, and a typical set of daily results for all five workers. In the spreadsheet itself, the 50 beads in the paddle will appear as either white or red. In Figure 1.10 the darker beads are the red ones.

You can use the spreadsheet to generate your own set of results, or simply use those shown in the table in Figure 1.10.

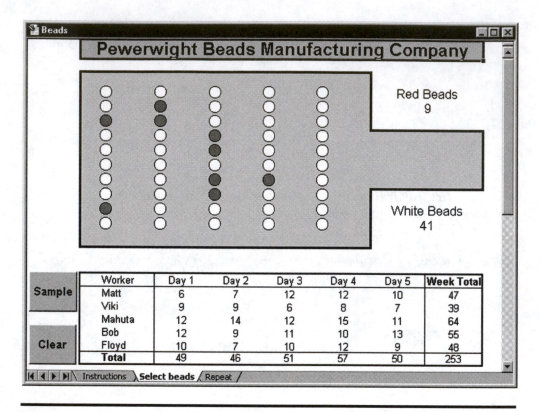

Figure 1.10   Beads Experiment Spreadsheet

1.9   Exercises

1. In the beads experiment there were big differences in performance between the five workers taking part. This was due to
   a.   a lack of motivation with some workers
   b.   inadequate training
   c.   chance
   d.   some workers not taking the exercise seriously
   e.   pressure from the supervisor
   f.   none of the above

2. The high level of defects in the beads experiment was caused by
   a.   workers not mixing the raw material properly
   b.   inadequate training of the workers
   c.   poor raw material
   d.   lack of an incentive scheme
   e.   insufficient supervision

3. In the beads experiment, if the number of red beads in the raw material was considerably reduced then there would be
   a.   more
   b.   same
   c.   less
   variability in the number of defects produced.

4. In the beads experiment, which of the following proportions of red beads in the raw material do you think would give the greatest variability in the results?
   a. 5% red
   b. 10% red
   c. 20% red
   d. 40% red
   e. 80% red

5. (B. Joiner: *Fourth Generation Management.*) A small company launched a new product that turned out to have amazing success. Though there were a few bad months, sales kept increasing month after month after month. The revenues this product brought in provided a much-needed relief to the company's line of credit and supported new expansion efforts. Then, suddenly, there was a downturn. There were three bad months in a row.

   How would you react to the dip in sales in this key product? Would you
   a. try to motivate your sales staff?
   b. reduce prices?
   c. place additional advertisements?
   d. mail promotional materials to potential new buyers?
   e. call previous buyers or prospects and see why sales had dropped?
   f. do nothing?

6. (P. R. Scholtes: *The Leader's Handbook.*) A psychologist attached to the Israeli Air Force had heard reports that the flight instructors were being abusive of the student pilots: there were no compliments but instead there were public reprimands, screaming, and abusive language. On further examination, he found out that the performance of the student pilots generally improved after they were reprimanded and, indeed, got worse after they were complimented. What the flight instructors claimed seemed to be true. Based on student pilot performance, he could no longer advocate using more "positive reinforcement" nor urge the flight instructors to avoid the harsher responses.

   The psychologist gave up this project and moved on to other work. However, the results of this study troubled him. During the following year he learned something that led him to reexamine the abusive flight instructor problem. When he interpreted the results with a different perspective he found that the flight instructor reprimand did not result in improved performance, nor did compliments result in a worse performance.

   What was the psychologist's newly acquired perspective?

7. In a fast food chain, management has introduced a "worker of the month" incentive scheme. Do you think this will motivate staff? Explain the reasons for your answer.

8. Jim, the customer services manager of a large courier company, was concerned about the number of mistakes being found in customer invoices. He collected some data on the type and frequency of mistakes. He found that the biggest problems related to the work of Mary, one of the data input clerks. Jim also discovered that Mary had been warned before about the number of mistakes she was making. Jim is now considering whether to terminate Mary's employment with the company. What other information should Jim consider in making this decision?

Chapter 2

Problem Solving

2.1   A Scientific Approach to Problem Solving

Problem solving in most organisations tends to be concentrated on "fire-fighting" obvious and urgent problems. We need to learn instead to:

- manage the organisation as a system,
- develop process thinking,
- base decisions on data, and
- understand variation.

In other words, we need to use a scientific approach. Knowledge about variation is a key element of the scientific approach. Our ability to solve problems quickly and in a sustainable way is directly linked to our ability to understand and interpret variation; otherwise any decisions made may well make things worse, or have no effect at all, rather than make things better.

Unfortunately, a great deal of management time is taken up with fire-fighting. Even worse, the problems "solved" this way often disappear only for a short time, reappearing in the same or a similar guise a few days, weeks, or months later, to be solved all over again. Many problems are competing for attention, but with fire-fighting only enough effort is made to put out the immediate fires. The problems keep coming back time and time again. It is rework. This "ready-fire-aim" approach is wasteful and ineffective.

The scientific approach means:

- making decisions based on data rather than hunches,
- looking for root causes of problems rather than reacting to superficial symptoms, and
- seeking permanent solutions rather than quick fixes.

The scientific approach leads to sustainable improvements in *all* areas of business such as cost, meeting delivery schedules, employee safety, skill development, supplier

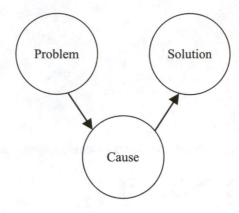

Figure 2.1   The Problem–Cause–Solution Approach

relations, new product development, or productivity. It is more effective if a number of people work together as a team to pool their knowledge and experience in order to solve problems.

Putting solutions in place, through a process of identifying and eliminating root causes, is a more effective approach to problem solving. This is depicted in Figure 2.1.

Fire-fighting omits the cause step. It is an instant solution approach that rarely works. The debt recovery example given in the previous chapter is such an example. The general manager thought he was being effective, but because he did not understand variability, and was not addressing a cause of the problem, the overall level of debt did not improve.

In this chapter we study the scientific approach, and use the debt recovery example to illustrate how this approach can be effectively used to solve problems.

2.2   The Need for Data

We have already stated that data are needed for decision making, and for understanding and interpreting variation. Data across all business operations are needed for informed decision making, such as data collected from sales records, inventory levels, accounts receivable, daily production records, and defect reports. In many organisations such collection is routine, although the purpose is often not clear and much of it is never used. It is wasteful to obtain more data than is needed to make the appropriate decisions.

For problem solving, some of the reasons for collecting data from a process are

- to *understand the process*. The system may not be well enough understood to be able to pinpoint the fundamental causes of the problem. Some basic data are needed in order to know what is going on in the process.
- to *determine priorities*. There may be a number of potential causes of the problem. It may, therefore, be necessary to make decisions about which causes should be examined immediately and which left to some future investigation. This can be achieved by collecting data on the process so that the potentially important causes can be identified.

- to *eliminate causes of variation*. For instance, can the overall level of defects be reduced or the variability in the results decreased? We shall see later that to do this it is important to distinguish between two types of causes, namely special and common causes.
- to *monitor the process*. It may be known what behaviour the process exhibited in the past. Does it still exhibit the same behaviour? Did the policy change instituted two months ago improve the process?
- to *establish relationships* between different factors or characteristics affecting the process. It may be, for instance, that the variable of interest is difficult or expensive to measure. However, a strong relationship may exist between this variable and another variable, which is easier or cheaper to measure. It may then be possible to use this substitute in subsequent analyses.

Debt Recovery Project

It has to be clear at the outset what is the purpose, scope, and aims of any problem-solving project. As stated in the previous chapter, the level of unrecovered debt has been a cause of concern for some time to the Albion Cleaning Company. The visits made by the general manager to the district offices have not been effective in reducing the overall level of debt. Even worse, they are costly and are contributing to a loss of morale, especially among the district managers. It is clear that this problem needs urgent attention. In particular:

- the high level of unrecovered debt is having a serious impact on the company's cash flow,
- lost interest alone on the $1,000,000 of unpaid invoices each month is not insignificant, and
- the problem is believed to be leading to numerous customer complaints, and to upset and lost customers.

An improvement team consisting of the general manager and other senior managers was set up to study this problem. Their aim was to determine the causes of the high level of unrecovered debt, to eliminate those causes, and to prevent them from recurring. They set the following goals:

1. in the short term, to produce results that are consistently above a target level for recovered debt of 80%; a figure already shown to be achievable, and
2. in the longer term, to achieve a level of debt recovery of 90%; a goal felt to be realistic.

Q1. *The purpose and reasons for choosing this project are clear. The goals are realistic and achievable. What do you think the team should do first?*

2.3   Deciding What to Measure

Deciding what to measure is often a very difficult problem. To collect meaningful data we must know precisely what to observe and how to measure it. Consider the following example from Joiner's *Fourth Generation Management* (McGraw-Hill, 1994).

One American airline had a very customer-focused goal of making sure flights had on-time departures. Their employees had this goal as one of their most important objectives. The airline put out advertisements to the effect that they were the best airline in terms of on-time departures. Its safety record, baggage handling, and other services were no worse than its competitors'.

> *Q2. Suppose you are a regular business traveller, flying most weeks to different cities within the U.S. Flight delays are a great irritation causing you to be late for meetings, or late getting home. Would you be likely to fly with this airline, and why?*
>
> *Q3. Brian Joiner used to fly regularly with this airline and had exalted frequent flyer status. He flew with them so often that he got a free, automatic upgrade to first class on every flight. But their on-time departure goal upset him so much that he flies with this airline now only when absolutely necessary. What do you think could have upset him?*

We also have to be very clear as to the purpose of collecting the data. Problems can occur when the measurements are inconsistent from day to day, or from person to person, or from machine to machine. Unless we get consistency, difficulties will arise between, for instance, customers and suppliers, or people within the same organisation. For example, a salesperson is told that her performance will be judged on the percentage change in this year's sales over last year's sales. What does this mean?

- Average percentage change each month? Each week? Each day?
- For each product?
- Based on sales volume or sales revenue? Gross sales? Net sales?
- Using last year's prices or this year's prices?

Often customers specify what they want in vague terms that are difficult to define and measure. For example, what is meant by:

- Good tasting ice-cream?
- Good or bad service in a restaurant?

- On-time delivery?
- Smoothness?
- Fresh?
- User-friendly?
- Easy to assemble?

Although it is frequently difficult to get precise definitions, we have to persevere. We have to figure out what measure, or range of measures, has a bearing on these vague customer requirements. We should aim to provide an *operational definition* that removes all ambiguity by describing what something is and how to measure it, and translating vague concepts into procedures that everyone can use.

Q4. How might you measure good or bad service in a restaurant?

2.4   Mapping the Process

Process flowcharts help us to understand a process by mapping out, step-by-step, in a diagrammatic form, a sequence of actions that take place in the process. They provide a means of arriving at a common understanding of what the current process does, an insight into where in the process poor quality, waste, delays, or inefficiencies might be occurring, and a good framework to show the effects of the problem-solving effort.

The simplest type of flowchart is the *top-down flowchart*. This maps out the major steps of the process, concentrating on the essentially useful activities that take place. Having identified these major steps, we can then add more detail of what is involved in each step. If attention has been focussed on a particular process for the purpose of problem solving, then a top-down flowchart is often all that is required to understand what is going on in that process. Where the flow of a process is more complex, or where more detail needs to be shown, other flowcharts may be necessary. *Decision flowcharts* split a process into different paths based on some decision being taken. *Deployment flowcharts* add further information about who or what department is responsible for each step of the process. *Work flowcharts* combine a plan of the physical layout of the process with indications of which route the work takes as it moves through the process. Scholtes, in *The Leader's Handbook* (McGraw-Hill, 1998), gives examples of these and other types of flowchart.

In the debt recovery project the senior management team did not have a clear picture of what was involved in generating accounts, issuing invoices, collecting payments, and recovering debt. To make sure that all members of the team had

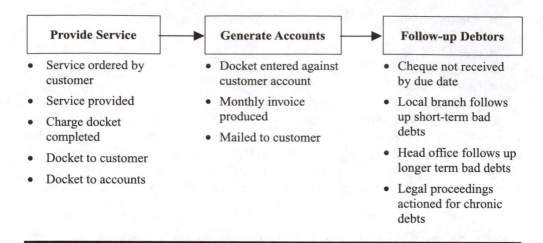

Figure 2.2   Top-Down Flowchart for Debt Recovery

the same understanding of what was involved, a top-down flowchart of the debt recovery process was constructed. This flowchart is given in Figure 2.2.

Q5. What would you do next if you were part of the debt collection project team?

2.5   Cause Analysis

The next step is to identify the potential causes of the problem. A list of the potential causes enables us to develop specific improvement strategies. The *cause and effect diagram*, also called a *fishbone diagram* or an *Ishikawa diagram*, is used to list the potential causes in an organized way. By systematically observing and recording these causes, the structure or relationship between cause and effect can be determined. A top-down flowchart of the process will be an invaluable aid in constructing a cause and effect diagram. The team working on the debt collection problem used brainstorming methods to produce the cause and effect diagram in Figure 2.3.

Examining Figure 2.3 you will see that it lays out the process as a convergence of activities or causes that result in the final event or effect. Major cause/activity lines converge on the central result line, and minor causes/activities that make up each major cause are plotted as short lines along the major lines. The diagram can "drill-down" to as many levels as required, although three cause levels is usual, as shown here.

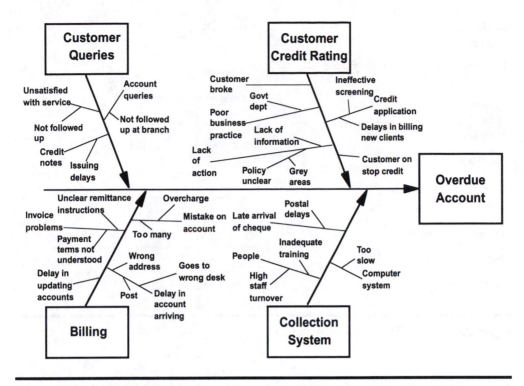

Figure 2.3   Cause and Effect Diagram for Debt Recovery Project

Q6. *How are you going to use the information given in the above cause and effect diagram? What should you do next?*

2.6   Collecting Data

There are very many possible causes for overdue accounts in the debt collection project. Which one should we address first? What are the most likely causes of the problem? The general manager thought that postal delays were the cause of the problem, and said they should use a courier company, instead of the postal service, to deliver the invoices to customers and to collect payments. How do we know this solution will work? We could try it and see, but this is what we do when we fire-fight a problem. If postal delays are not the cause of the problem, then we shall have put in place a solution that not only costs more but also does not lead to improvement. It may also be some months before this solution is shown to be ineffective.

To avoid addressing just the symptoms of the problem, it is necessary to identify and verify the root causes. It is impractical to collect data on all the possible causes in the cause and effect diagram, so some prioritising is necessary.

DATE: *Thursday, 5 October 1992*

COLLECTED BY: *Jim Brown*

NUMBER OF PAYMENTS RECEIVED: *130*

Area Posted	Number of days to deliver payment				
	1	2	3	4	5+
Northland	|||				
Auckland	卌 卌 卌 卌 卌 卌 卌 卌 卌 卌 卌 卌 卌 卌	卌					
Waikato	卌						
BOP / Hawkes Bay	卌 |||	||||			
Taranaki / Manuwatu		卌 ||||	||||		
Wellington	||	卌 |||					
South Island				||||			

Figure 2.4   Check Sheet for Postal Delays

The team identified four possible causes that should be looked at first: postal delays, account queries, mistakes on account, and government departments. They decided first to collect some data on how long it took for customer payments to reach the Auckland office, where payments are processed.

Care has to be taken in collecting data. Check sheets are one of the primary tools for gathering data. They are structured forms that make it easy to collect data. A well-designed check sheet improves the reliability of the information collected. It also builds a picture of what is going on in the data, thus simplifying further analysis. An example of a check sheet is shown in Figure 2.4.

The following guidelines will help you make good check sheets and gather good data:

- Use *operational definitions*. Operational definitions remove ambiguity from check sheets, as they define what is to be measured and how it is to be measured.
- Incorporate a *visual element* if possible. For example, suppose we are studying damage on cardboard cartons. The check sheet could consist of a picture of a carton, and the position of the defect can then be noted on this picture.
- Include a section for recording *background information*, such as date, time, operator, machine, and so on. Without this information it will be difficult to undertake a full analysis of the data later. For example, did the results differ from machine to machine?
- *Test* your check sheet. Use the check sheet to collect, and then analyse, a small set of data. Did it work as well as you thought? Based on what you learned, improve the check sheet, and then begin your data collection effort.

- *Design* it so that questions are answered from the top to the bottom of the page. Limit it to a single page, and allow extra room for comments.
- Keep asking yourself, "Can it be improved?"

Debt Recovery Project—Data on Postal Delays

The check sheet in Figure 2.4 was used to collect data on postal delays. The survey of mail was carried out on different days over a number of weeks. The results of one of the days are given in the check sheet. From the postmark, the number of days taken for the mail to reach the processing office in Auckland was recorded. The areas listed are in order from the north to the south of New Zealand.

A good picture of the delivery times is built up as the data are collected on the check sheet. A bar chart of these data is given in Figure 2.5. Similar results were obtained on the other days of the survey.

Q7. *What do you conclude from the above analysis? What should you do next?*

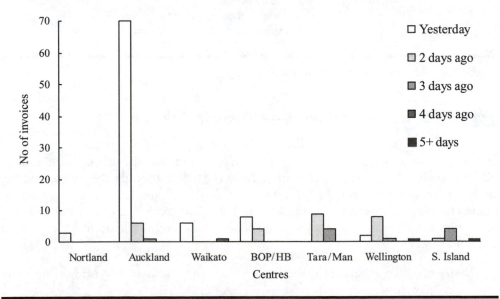

Figure 2.5   Bar Chart of Postal Delays (Thursday, 5 October 1992)

Debt Recovery Project—Data on Account Queries

The debt collection team concluded from the survey of incoming mail that postal delays were not a cause of the problem of overdue accounts. They revisited the cause and effect diagram, and decided to collect information on account queries. The data obtained are plotted as a Pareto diagram in Figure 2.6. Pareto diagrams will be discussed in more detail in Section 3.8, but the figure is basically a bar chart with the main reasons ordered by magnitude.

Q8. What do you conclude from Figure 2.6?

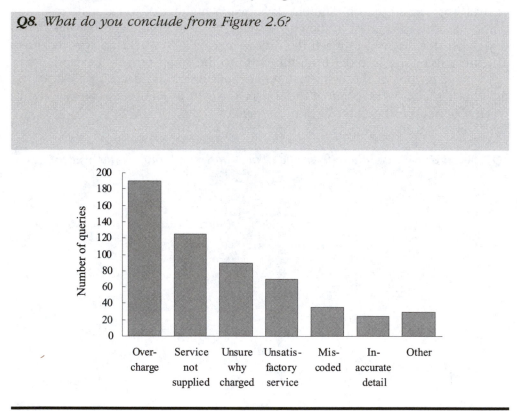

Figure 2.6   Account Queries (November 1992)

Debt Recovery Project—A Possible Solution

It was clear to the team that there was a very large number of account queries, and that customers were not paying their invoices because they thought they were being overcharged, or were not getting the service they thought they were getting, or were unsure what service they were getting. Various aspects of the service contracts were ambiguous and unclear, and the information given in the invoices was uninformative and incomplete. The actions taken by the team to reduce these difficulties were

- to produce a brochure explaining the range of services offered and the cost of providing each type of service, and
- to redesign the invoice form so that it is clear to the customer what service has been provided and the cost of that service.

Q9. *What should you do next?*

2.7   Evaluating the Results

The debt collection project team has now identified a cause of the problem that has been verified by data. They have taken action to remedy the situation. However, we cannot assume that the action taken will lead to an improvement in the results. We now need to evaluate the actions by comparing data on the process before and after improvement. For instance, in another project aimed at reducing the amount of damage on cardboard packaging, data clearly showed that the young, inexperienced forklift operators were causing more damage than the older, experienced operators. As a result it was decided to put the young operators through an extensive retraining program. Subsequent data revealed, however, that this solution had no effect. Further investigation led the project team to discover that the new operators were driving old forklifts that were no longer suitable. Once modifications were made to these forklifts, the differences between the operators were eliminated.

In the debt collection project the company continued to collect data on the percentage of debt recovered each month. The data up until May 1993 are shown in Figure 2.7.

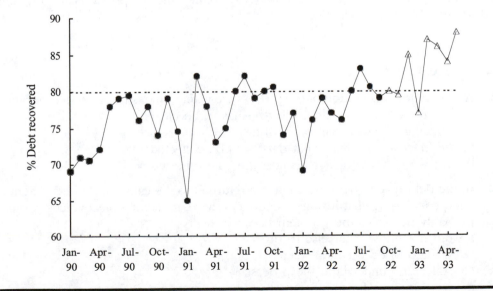

Figure 2.7   Debts Collected Within Due Month

> **Q10.** *Is there any evidence to suggest that the solutions put in place have been effective? In particular, what can you say about the January 1993 result?*

2.8   Standardising and Planning for Future Improvements

To ensure the solution is fixed in place, the team took the following actions:

- the brochure was issued to each salesperson and became the basis for negotiations with present and new customers,
- the new invoice form was adopted as the standard form; all previous forms were destroyed, and
- a copy of the improvement story was sent to all district managers.

In reviewing what had been achieved, they concluded that improvements made had met the short-term goal of reducing the percentage of unrecovered debt to less than 20%. However, the team believed that still more could be done and so they decided to continue working on this project. To reduce the percentage of unrecovered debt further, they looked again at the cause and effect diagram. As a result they decided to focus on those customers who were slow to pay. They suspected that this group included government departments.

Through understanding the nature of variability, the use of data and applying a scientific approach to problem solving, the senior management team had made significant improvements in just a few months. The percentage of unrecovered debt had been a concern for a number of years, and all previous fire-fighting efforts had been unproductive.

2.9   Exercises

1. An operational definition of the process we want to measure is needed because
   a. the reason for collecting the data has to be clear
   b. errors in the data will be hard to detect otherwise
   c. we need to be clear what data is to be collected
   d. we won't know how to collect the data otherwise

2. In the debt recovery project the team returned to the cause and effect diagram after concluding that using a courier service to send out invoices would probably not solve the problem because they
   a. did not know what else to do
   b. wanted to collect more data
   c. still didn't understand the process
   d. were looking for another reason for nonpayment
   e. needed to further explore the problem

3. A project team set up to study the problem of damage to packaging cartons found that the young forklift operators were causing more damage than the older, experienced operators. Providing further training for the young operators in the use of forklifts was not effective. This was probably because
   a. driving a forklift is an unskilled job
   b. they were not concerned about carton damage
   c. they already knew how to drive forklifts
   d. inadequate training was not the cause of the problem
   e. they were not causing much damage anyway

4. In the debt collection project, the team went through one cycle of the scientific approach.
   a. What lessons can the team learn from what they achieved?
   b. Can you think of any ways of improving what the team did?

5. A whiteware manufacturer is concerned about the occurrence of visible scratches, blemishes, and other surface imperfections on their refrigerators. Management first ordered a motivation programme to persuade workers of the importance of refrigerator appearance. Next, the quality of the paint was improved. Neither actions helped; so, it was decided to add a special "touch-up" operation at the end of the production line to correct any surface defects.
   a. Critically comment on the actions taken by management.
   b. Can you suggest an alternative approach to the problem?

6. Explain the purposes of check sheets. Construct a check sheet to study the reasons for absenteeism in a factory.

7. A major hospital is concerned about the length of time required to get a patient from the emergency department to an in-patient bed. Significant delays appear to be caused by beds not being available. What would be the first steps you would take to tackle this problem?

8. Mary Thompson, the general manager of an up-market hotel, is concerned by the number of customer complaints received over the last three months. What are the first steps Mary should take to address this issue?

Chapter 3

Looking at Data

3.1 Types of Data

At first sight, it is tempting to think that data are just numbers that describe some process or situation. However, consider the following three pieces of information that describe a particular company:

- it is an insurance company,
- its financial security rating is Aa,
- in the last financial year, its turnover was $756.3 mil.

Each of these can be considered as a piece of "data," but only the third one is expressed numerically. The first two are examples of *qualitative* data, whereas in the third example the data is *quantitative*.

↯ Qualitative Data

Within the category of qualitative data, it is usual to distinguish between *attribute* (or *nominal*) data and *ranked* (or *ordinal*) data. Attribute data simply describes a characteristic (attribute) that some item or individual possesses, which either serves to distinguish one item from another, or classifies each item into one of various mutually exclusive (and usually exhaustive) categories. The first example above is attribute data that describes the activity of the company, and distinguishes it from other types of business such as manufacturing or distribution. Other examples of attribute data are

- A player's number in a football team. Note that numbers here are only used conventionally. They could equally well be letters or the players' names, which are sometimes used as well as (or instead of) numbers.
- A list of the different types of mistakes made on a particular job.

- Absences from work. These could be categorised into *sickness, injury, leave with pay, leave without pay, incapacity of a close relative, other*. The final category is usually needed to cope with rare, infrequent reasons and with situations we may not have thought about in advance.
- Immigration forms at airports. These ask for your marital status using categories such as *never married, now married, separated, divorced, widowed*.
- Those where there are just two categories (the simplest type of attribute data). For instance, *defective* and *satisfactory, good* and *bad, pass* and *fail*, or *yes* and *no*.

The categories or labels used are often expressed as numbers. For example 0 = male, 1 = female. This should not disguise the fact that such data are not truly numeric, and so arithmetic is not appropriate. For example, it is not sensible to add up a set of such 0's and 1's to give an "average" sex. The only property possessed by attribute data is that of equivalence. For example, if two people each have value 0 they are the same sex (male).

As soon as order is implied between individuals or categories then the data are *ranked*, or ordinal. It may be that one category is preferred to, or is better than, or is higher than another category. In whatever way, the individuals or categories can be placed in an implicit order. The second example in the previous section is ranked data in that the categories used (Aaa, Aa, A, Baa, etc.) are in decreasing order of financial security. In some cases, all the items or individuals forming a sample of data are ranked from first to last in some way, such as the results of a horse race, with few or no ties. More often, each item is placed into one of a number of ranked categories, such as the grades you get on a course (A+, A, A−, B+, B, …), with items falling in the same category being seen as equivalent, but higher or lower than another category.

This latter kind of data often arises in market research where respondents are asked to express their opinions about something on a five-point scale, such as *strongly agree, agree, indifferent, disagree*, and *strongly disagree*. It is important to note that, although one category is higher or lower than another, it is not possible to infer how much higher or lower. In particular, the "difference" between *strongly agree* and *agree* may not be the same as that between *agree* and *indifferent*, in the same way that the difference (in marks) between the first and second placed students may not be the same as that between second and third.

When dealing with ranked data, only calculations involving relative position are meaningful. For example, it is possible to calculate an average of ranked data using the median, but not using the arithmetic mean. The median is defined in Section 3.11, and the arithmetic mean in Section 4.3.

Quantitative Data

Once it is possible to say by how much one observation is greater than another, the scale of measurement is called *interval*. Order exists again but the magnitude between values has meaning and provides extra information. Turnover of a company is interval data, as are the following examples:

- the time taken by each horse to complete a race,
- the number of beers consumed weekly by students at the University of Waikato,

- figures that describe the state of some process, such as the amount of an additive, the temperature of an oven, the rotational speed of a motor, or the weight of packages produced.

It is possible to distinguish different types of quantitative data (for example, between *interval* and *ratio* data) but this will not be relevant here, as most simple statistical procedures are applicable to all forms of quantitative data.

Often only attribute data are available, but greater information is obtained with ranked and, especially, interval data. The aim should always be to strive to obtain interval data, although this is not always possible. Avoid throwing away information by converting interval data to attribute data. For instance, the lengths of some critical component are measured but the only information recorded is whether these lengths are within or outside some specification limits. They could all be within the limits; so, the attribute data say nothing, whereas an important difference could be indicated by the interval data.

Q1. List three additional examples of attribute and interval data.

3.2   Presentation of Data

Particular attention should be given to the presentation of numerical results. Badly presented data in tables or graphs can make any report ineffective. It can also mislead. Consider the following example.

   Petrol Consumption Example

The purpose of the data in Figure 3.1 is to provide a comparison of car petrol consumption (km/litre) at constant speeds (km/hr) for five different cars A, B, C, D, and E.

Car	Speed (km/hr)			
	50	65	100	160
A	19.09	13.20	10.09	7.27
B	20.06	17.43	14.65	9.06
C	14.06	14.57	17.42	13.58
D	10.24	14.93	12.33	5.13
E	11.55	14.58	13.34	6.23

Figure 3.1   Car Petrol Consumption (km/litre)

Q2. *Do you find this table easy to read at a glance? List some of the things that make it hard or that distract you from making a useful summary of the information contained in this table.*

A number of suggestions for more effective presentation will now be considered. The books by Ehrenberg (*A Primer in Data Reduction*; John Wiley & Sons, 1982) and Tufte (*The Visual Display of Quantitative Information*; Graphics Press, 1983) are the basis for these suggestions.

3.3   Rounding Data

One major problem with reading numerical data is that it is difficult to conceptualise and/or mentally manipulate long numbers. A possible way of overcoming this difficulty is to round each number to *two effective digits*.

It is common practice to show percentages to three significant digits, such as 17.9% and 35.2%, but with numbers such as these it is difficult to mentally divide the smaller into the larger, or even to subtract the smaller from the larger. Rounding to two effective digits gives 18% and 35%. It is now easy to see that one is just under twice as large as the other, and that the difference between them is about 17%.

Consider index numbers like 117.9, 135.2, 128.6, and 144.3. The initial 1's are not "effective," since they do not vary in these data. They are "dead" digits. The other two digits are the effective or "busy" digits. Therefore, the data to two effective digits are 118, 135, 129, and 144. These numbers are still quite easy to manipulate because the initial 1's are the same. However, if the next index number is 93.3 rounding to 2 effective digits would give 90, 120, 140, 130, and 140, which is far too great a discrepancy. Now it is better to use three effective digits to give 93, 118, 135, 129, and 144. The basic idea is to use as few digits as possible, for ease of comprehension, while introducing as little distortion as possible by the rounding.

When there is wide variation in data, a variable standard for rounding can be used. For example, values such as 223.3 and 34.7 would be rounded to 220 and 35, respectively. This variable rounding means that the rounding does not really affect any comparisons between the numbers.

The information lost by dropping the other digits is usually negligible, yet the advantage in terms of communication is considerable. It is not suggested that the original data be discarded. It may be that the other digits are important in, for instance, monetary terms. The rounded data are simply to be used in reporting the results in order to add clarity to the report.

3.4   Effective Tables

Data are very often presented in tabular form. There are several simple principles which can be followed in setting out a table, and which make the table easier to follow.

1. *Round the results to two effective digits*, as described in the previous section.
2. Use *averages* to indicate the overall structure of the table. Row and column averages can provide a focus to guide the eye when looking at the individual figures. Even when the pattern varies from row to row, or from column to column, the row and column averages provide a useful focus to help see the nature of the differences.
3. Where possible, *order rows and columns by size*. This helps to identify patterns and exceptions to the data.
4. *Numbers are easier to read down than across*. This is because the leading digits in each number are then close to each other for direct comparison, with no other digits in between. Thus the table should be arranged with the categories whose comparison is of greatest interest down the left-hand side of the table.
5. Have a *simple layout* that is easy to read. Table layout should make it easy to compare the relevant figures (the two preceding guidelines are also based on this criterion). The primary reason for the existence of the table is to convey information, not simply to impress the reader. In particular:
   - too many gridlines, widely spaced rows, and irregular spacing of rows and columns impede visual comparison and can be distracting,
   - the intelligent use of white space in the layout can be used to convey the underlying groupings and emphasise any patterns in the data,
   - reducing the size of a table can also be useful. It also reduces the amount of eye-movement required to scan the table,
   - all tables should, of course, be clearly labeled.
6. A brief *verbal summary* helps to focus attention on the salient features of the table, and consequently makes the whole report so much easier to read and understand.

Petrol Consumption Example

Q3. Using the guidelines given above, set out the data given in Figure 3.1 in a more easily comprehensible table.

Q4. Write a sentence or two to summarise the information contained in these data.

3.5 Graphs

Graphs are used essentially to convey information about relationships, shapes, relative sizes, and priorities, rather than to present strict numerical information. They help to make large sets of data coherent. There are a few things to bear in mind when using graphs.

- If a complex story is to be illustrated graphically, then it is better to do so in a series of graphs, each of which illustrates a single element of the story.
- Graphs can reveal the data at several levels of detail, from the broad overview to the fine structure.
- It is important that all aspects of the graph are clearly labelled. The graph should have a heading that gives sufficient information to identify the original data. The axes should have titles and the scales should be clearly marked.
- Over elaboration should be avoided when constructing graphs. The substance of the graph should be the centre of attraction, not the design graphic, the technology of graph production, or something else. Over elaboration often leads to a distortion of the information contained in the data.

In the following sections, some of the more commonly used graphs will be described and illustrated.

3.6 Bar Charts

The simplest form of graph is the bar chart, which can be used for attribute, ranked, or interval data. We have already seen an example of a bar chart in Figure 2.5, where it was used to depict postal delivery times.

The check sheet in Figure 3.2 records the reasons for the issue of 212 credit notes during a 2-month period. A list of reasons is drawn up for the issue of credit notes and, as each note is checked, the reason for its issue is tallied off. A good picture of the patterns of credit note issue is built up.

Q5. Why would the company collect such data?

Most graphs, including bar charts, are very easily created in Excel, as described in Appendix A, Section A.9. First the data in Figure 3.2 must be entered into a spreadsheet as in Figure 3.3. Using the chart wizard, follow steps 1 to 4 selecting a column chart, with data in A2:B9 arranged by columns. The X and Y axis titles are input in step 3, and the legend and gridlines suppressed. Pressing *Finish* creates the required bar chart on the spreadsheet as shown in Figure 3.3. Each of the reasons (categories) identified in Figure 3.2 is represented by a bar whose height is equal to the number of times that reason occurred.

LOCATION:	*Alphington*	DATE INSPECTED:	*Nov 1989*
PERIOD COVERED:	*Sept / Oct*	INSPECTOR:	*Alf James*
NO OF INVOICES ISSUED IN PERIOD:	*3797*		

REASON	TALLY	
Damaged Goods	‖	2
Incorrect Goods	‖‖ ‖‖ ‖‖ ‖‖ ‖‖ ‖‖		31
Nondelivery	‖‖ ‖‖ ‖‖ ‖‖ ‖‖	24
Not Required	‖‖	3
Out-of-Stock	‖‖ ‖‖ ‖‖		16
Pricing Error	‖‖ ‖	108
Shortage	‖‖ ‖‖ ‖‖ ‖‖ ‖	22
Other	‖‖ ‖	6
TOTAL		212

Figure 3.2   Check Sheet for Credit Notes

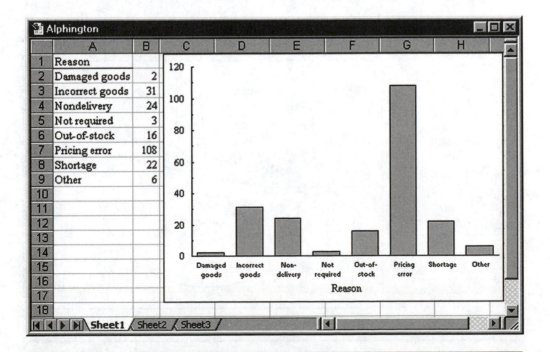

Figure 3.3   Bar Chart of Alphington Credit Notes

Q6. What conclusions do you draw from this chart?

3.7   Pie Charts

From Figure 3.3, it is quite clear that pricing errors are the most frequently occurring reason for the issue of credit notes, and that damaged goods or goods not required occur very infrequently. But what proportion of all credit notes is due to pricing errors?

Q7. Without performing any detailed calculations, make a quick estimate of this proportion based on the bar chart presented in Figure 3.3.

To make even a rough estimate of this proportion is not easy. This is because Figure 3.3 does not show the information in the most suitable form. When we are primarily interested in the proportion of the total that is accounted for by each category, it is better to present the information as a *pie chart* in which each of the categories is represented by appropriately sized segments of a circle. A pie chart for the Alphington credit note data is shown in Figure 3.4.

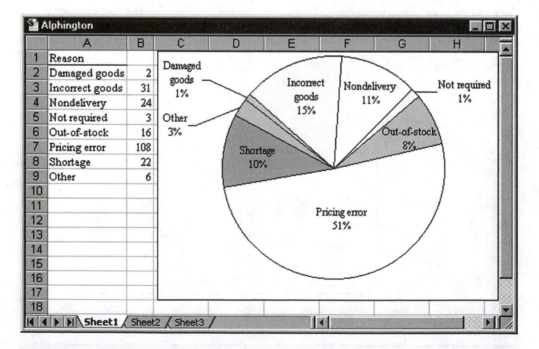

Figure 3.4   Pie Chart of Alphington Credit Notes

The pie chart is created by almost exactly the same procedure as the bar chart in Figure 3.3. The only difference is that chart type *Pie* is selected at step 1, and *Data Labels* (instead of *Legend*) is selected at step 3. This allows the percentage split for each category to be included on the chart as shown.

It is now clear, even if the percentages were not given on the pie chart, that about 50% of all credit notes are due to pricing errors.

3.8   Pareto Diagrams

The Pareto diagram is a bar chart that ranks problems (categories) by how frequently they occur. Each block represents a problem; the bigger the block, the bigger the problem. It helps to identify which problems need immediate attention and which can be looked at later. While there are many factors that cause a large problem, it is frequently found that a few specific factors account for the bulk of the observed occurrences of that problem. The Pareto principle can be depicted as in Figure 3.5.

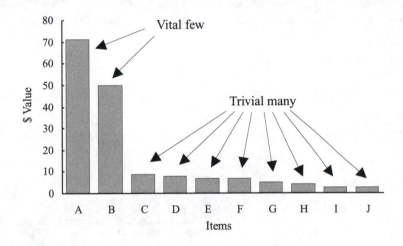

Figure 3.5   The Pareto Principle

It has been shown many times in practice that a lot more progress can be made if problem-solving efforts are concentrated on the main causes of a problem rather than trying to tackle a bunch of lesser causes. Solving the biggest problem will usually result in the greatest improvement in the shortest amount of time. If attempts are made to solve all problems at the same time then some progress will probably be made, but it will usually be slower, and may even be counter-productive. Once the largest problems have been attacked then attention can be directed to the lesser ones, if this is warranted in terms of time and money.

 Credit Note Data

In the bar chart of the Alphington credit note data given in Figure 3.3, the different reasons, with the exception of *Other*, were plotted in alphabetical order. This is one possible ordering, but others could be chosen, as the order of categories does not matter. The Pareto diagram, setting out the categories in order of importance, is given in Figure 3.6.

As a Pareto diagram is no more than an ordered bar chart, the procedure in Excel is identical to that for the bar chart, provided that the data are first arranged in descending order in column B. Since the *Other* category comprises a number of other reasons, it is usually placed last in the diagram.

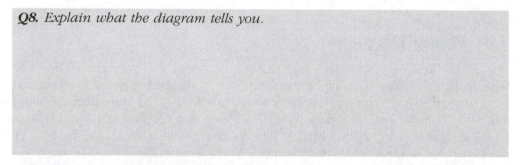

Q8. Explain what the diagram tells you.

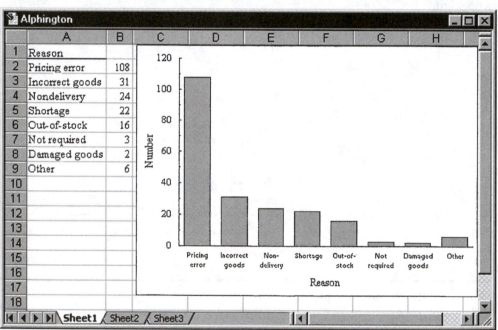

Figure 3.6   Pareto Chart of Alphington Credit Notes

The raw frequencies of occurrence may not be the best measure to plot on the Pareto diagram. If the value of each category is different, something more complex might be used. For example, looking at warranty failures of refrigerators, a broken plastic component costs much less to put right than a failed compressor. Thus the total costs of different categories of failure might be recorded, rather than simply the number of claims in each category.

The Pareto diagram is much easier to understand at a glance than a check sheet. For this reason they are often used to display data for a presentation, or to have on display in a department or on the shop floor, so that everyone can see where the problems are and seek corrective action.

❧ *Evaluating Improvement*

Pareto diagrams also enable improvement to be evaluated. Having identified potential major causes of a problem and taken corrective action, another diagram can be used to see how much improvement has taken place. If effective measures have been taken, then the order of items along the horizontal axis will usually change. In one problem-solving exercise, a company was able to reduce the amount of damage occurring on the sides of their cardboard cartons by replacing hooked clamps on their forklift trucks with regular clamps. The Pareto diagrams in Figure 3.7 show very clearly the effect of this solution.

> **Q9.** *What was the main effect of the change? Did it affect all types of damage found on cartons?*

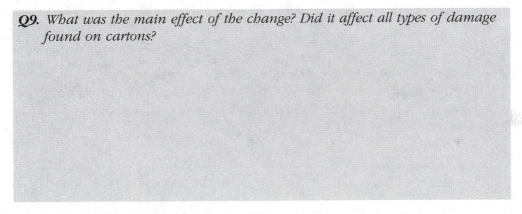

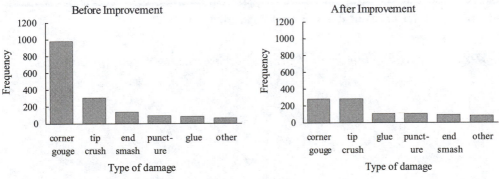

Figure 3.7   Pareto Diagrams Before and After Improvement

3.9   Run Charts

When data are collected from a process over some time period, it is often important to plot the data in the order in which it arose. Plotting the time sequence may reveal, for instance, the existence of some trends that would not be indicated simply from an examination of a bar chart or stem and leaf diagram (see Section 3.10). Alternatively, there may be changes in the average level at certain times, or changes in variability may be detected. A graph where data are plotted in the order in which they are obtained is called a run chart.

Various examples of run charts have been given in Chapter 1. An example is Figure 1.2 where the percentage of debt recovered is plotted over time. This chart was drawn in Excel by first entering the data into the spreadsheet with the months in column A and the percent debt recovered in column B. Selecting chart type *Line* at step 1 of the chart wizard, defining the data range (in columns) as A1:B33, and entering the required titles in step 3 gives the run chart of Figure 1.2, again having suppressed the chart legend and gridlines.

 ## Road Fatalities

The numbers of people killed on New Zealand roads, per 100,000 population and per 10,000 vehicles, for each year from 1951 to 1987 are given in the run charts in Figure 3.8.

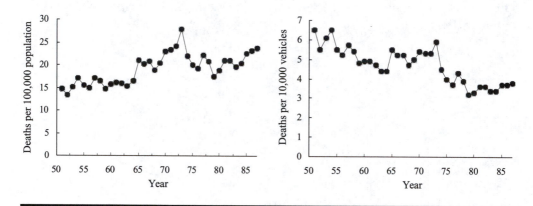

Figure 3.8   Fatal Accident Rates in New Zealand, 1951–1987

Q10. *What are the important features of these charts? What questions would you ask?*

 ## Aerofoil Effectiveness

The distribution centre of a large New Zealand manufacturing company was considering whether to fit streamlining aerofoils to their large trucks. The aerofoils are meant to give improved fuel consumption, but they cost $1000 each. In order to examine their cost effectiveness, the company decided to try it on one of their trucks. They chose a "B" train truck and trailer unit that made a daily run from Auckland to towns in the central North Island, a round-trip journey that usually varied between 300 and 400 km.

The trial started on June 29 without the aerofoil. The aerofoil was fitted on August 15, removed on October 16, and refitted on November 21. The trial finished on December 13. A run chart of the fuel consumption is given in Figure 3.9.

Q11. *What conclusions can you draw from this chart?*

Q12. *Are there any surprising results? If so, what would you do about them?*

Q13. *What other data would you like to see, or calculate?*

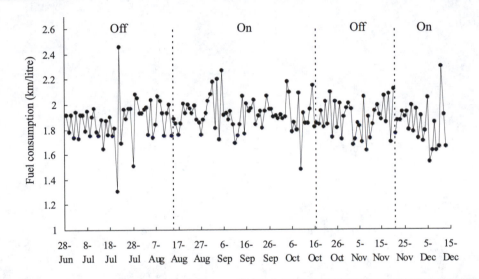

Figure 3.9   Daily Fuel Consumption

3.10   Stem and Leaf Diagrams

When a large amount of interval data has been collected, it is important to get an overall picture of it. A useful graphical aid for looking at interval data, which combines the features of a table and a bar chart, is the stem and leaf diagram.

The data are rounded to two effective digits, with

- the first effective digit used to form the stem, and
- the second effective digit forming the leaves.

Other digits can be ignored, as they will make virtually no difference to the picture obtained.

Consider the data in Figure 3.10, ranging from 115.6 to 204.8.

133.0	159.5	163.6	156.6	164.7	178.2	191.2	156.2
138.7	115.6	204.8	146.1	169.1	167.7	137.4	183.3
154.7	161.3	151.1	174.5	148.2	160.4	145.7	175.2

Figure 3.10   Sample Data for Stem and Leaf Diagram

As the hundreds digit is the same in all but one number, the two effective digits are the tens and the units. Draw a line down the page, and list the stems on the left side with one stem on each line in ascending order. Any ineffective leading digits (the hundreds in this case) are included as part of the stem. Put the leaves next to the appropriate stem, with one leaf for each measurement. Redraw the plot, rearranging the numbers on each stem in ascending order. The resulting stem and leaf diagram for the above data is shown in Figure 3.11.

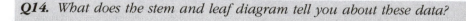

Q14. *What does the stem and leaf diagram tell you about these data?*

	Unordered		Ordered
11	5	11	5
2		2	
3	3 8 7	3	3 7 8
4	6 8 5	4	5 6 8
5	9 6 6 4 1	5	1 4 6 6 9
6	3 4 9 7 1 0	6	0 1 3 4 7 9
7	8 4 5	7	4 5 8
8	3	8	3
9	1	9	1
20	4	20	4

Figure 3.11   Stem and Leaf Diagrams

 Use of Stem and Leaf Diagrams

The overall shape of the stem and leaf diagram provides valuable information with which to assess and compare data.

- Is the plot symmetric or skewed? It is usually easier to work with and compare symmetric shapes. With skewed data it is sometimes desirable to transform the data by taking, for example, logarithms or square roots before carrying out any detailed analysis.
- Does the plot have a single peak, or two or more peaks? Two or more peaks may indicate that the data are the mixed output from two or more differing processes. Perhaps this is important. It does not necessarily indicate this, but it is pointing out a possibility.
- Are one or two data values very different from the others? We call such values *outliers*. They could indicate that something special was going on

with the process when these outliers were collected. On the other hand, such values could arise through some mistake being made in collecting or recording the data.

■ Are there clusters of data points or does the plot drop off suddenly at either end? These may indicate the presence of a number of processes at work. Or it could be that items have been inspected and those at the extremes discarded.

🏭 *Loughborough Savings Bank*

Loughborough Savings Bank has recently been reviewing the time it takes to process applications for a loan. The bank's target is to contact the applicant within three days of the initial application. William Hicks, the customer service manager, surveyed last month's loan records and extracted the data in Figure 3.12 on the number of days taken to process 37 loan applications.

2	3	3	2	4	5	4	5	4	3	3	2	2
4	1	2	2	1	3	3	7	2	3	4	3	3
4	2	1	4	2	5	3	3	2	3	5		

Figure 3.12   Number of Days to Process Loan Applications

A stem and leaf diagram of these data is given in Figure 3.13. Because the data only contain single digit values, there are no distinct leaves in this case, and zeros have been used. Any other character could have been used, such as an ×, to simply show the shape of the distribution.

Q15. *What percentage of loans is not meeting the standard of being processed within three days of the initial application?*

```
1 | 0 0 0
2 | 0 0 0 0 0 0 0 0 0 0
3 | 0 0 0 0 0 0 0 0 0 0 0 0 0
4 | 0 0 0 0 0 0 0
5 | 0 0 0 0
6 |
7 | 0
```

Figure 3.13   Stem and Leaf Diagram for Loughborough Savings Bank

Aerofoil Effectiveness

The stem and leaf diagrams for the aerofoil data are given in Figure 3.14. The left diagram gives the kilometres per litre when the aerofoil is fitted to the trailer unit, and the right diagram is when the aerofoil is off. In order to give a more detailed picture of the shape of the data, the stems have each been divided into two parts corresponding to the low (0 to 4) and high (5 to 9) leaves, respectively.

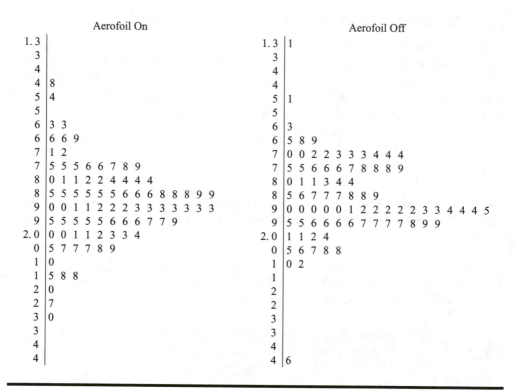

Figure 3.14   Petrol Consumption (km/litre) With and Without Aerofoil

Q16. What do you conclude from these diagrams?

Q17. How might you summarise the information contained in these diagrams?

3.11   Median, Range, and Quartiles

Consider again the (ordered) stem and leaf diagram in Figure 3.11. The 24 data values are set out in order of magnitude. The 12th value is 159 and the 13th value is 160. There are 12 values less than or equal to 159, and 12 values greater than or equal to 160. To find the value such that we have an equal number of data values above and below this value, we need to take the midpoint of the 12th and the 13th value.

The value that divides the data in half is called the *median*. It is a measure of the location or average of a distribution. The median value for the data in Figure 3.11 is 159.5.

The median is the value of the middle ranking observation, so that if the number of observations (n) is odd the median is given by the value of the $(n + 1)/2$ observation when the data are written in order of magnitude. For example, if $n = 23$ then the median is the 12th observation; 11 observations will be less than this value and 11 greater. If n is even then the median is halfway between the two middle observations, that is, the midpoint of the n/2 and next largest observations.

Now consider splitting the smaller half of the data in half again. In the example in Figure 3.11 there are 12 observations smaller than the median. To find the value such that we have 25% of data values below it and 75% above it, we need to take the median of the bottom half of the data, that is, the midpoint of the 6th and 7th observations. This observation is called the *lower quartile*. For our data, the lower quartile is (146 + 148)/2 = 147.

Similarly, we can split the top half of the data in half; this leads to *the upper quartile*. In the above example it is given by the midpoint of the 6th and 7th observations, counting backwards from the largest value. Thus, the upper quartile is (169 + 174)/2 = 171.5.

The median and quartiles are available in Excel as functions *MEDIAN* and *QUARTILE*. Specifically, the *QUARTILE* function returns one of five different quantities, namely the *minimum, lower quartile, median, upper quartile*, or *maximum* of a specified array, depending on whether the value of the second parameter (*Quart*) is 0, 1, 2, 3, or 4. This is illustrated in the dialogue box in Figure 3.15, which shows the calculation of the lower quartile (*Quart* = 1) for the data of Figure 3.11, located in cells A1:A24.

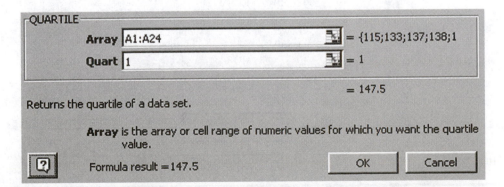

Figure 3.15   Quartile Dialogue Box

It should be noted, however, that Excel uses a slightly different definition of the lower and upper quartiles from that described above. In particular, Excel calculates the lower quartile of 24 observations not as the value midway between the 6th and 7th, but as the point three-quarters of the way between the 6th and 7th observations. Likewise, the upper quartile is taken to be a quarter of the way between the 18th and 19th observations. In practice, these small differences are not important in giving an overall picture of the data.

The smallest value is 115 and the largest value 204. The difference between the smallest value and the largest value is called the *range* of the data, and the difference between the lower and upper quartiles is called the *interquartile range* (IQR). The interquartile range is the range within which the middle half of the data lies. The range of the data in Figure 3.11 is therefore 89, and the interquartile range is 24.5.

3.12   Box Plots

We now have a number of convenient summaries of our data, namely:

Smallest observation = 115		Largest observation = 204	
Lower quartile = 147	Median = 159.5	Upper quartile = 171.5	
Interquartile range = 24.5		Range = 89	

Such information can be presented graphically with the *box plot*, which consists of a rectangle (*box*) covering the interquartile range, with the ends of the box corresponding to the two quartiles. The box is divided by a vertical line at the median. Joined to each end of the box are horizontal lines (*whiskers*), which extend outwards to the smallest and largest values in the data.

The box plot for the data in Figure 3.11 is shown in Figure 3.16.

The box plot is a useful graphical method for picturing the shape or distribution of a set of data. It is particularly useful when comparing different groups of

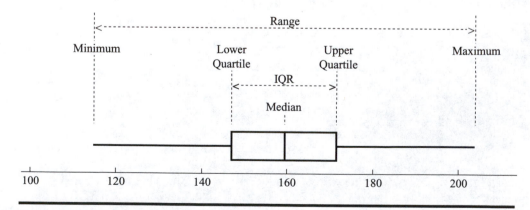

Figure 3.16   A Typical Box Plot

observations, by constructing a box plot for each group (on the same scale). A comparison of the plots will tell you about the symmetry of the distributions, as well as the centre and spread.

🏭 *Aerofoil Example Revisited*

For the aerofoil data, shown as a run chart in Figure 3.9 and a stem and leaf diagram in Figure 3.14, we can construct the box plots shown in Figure 3.17.

Q18. Why do you think asterisks have been used in these box plots?

Q19. What conclusion would you draw?

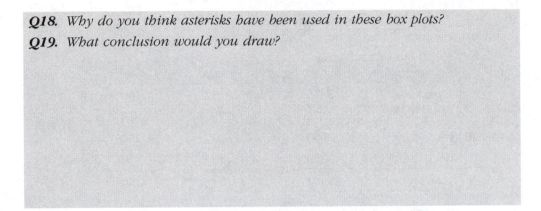

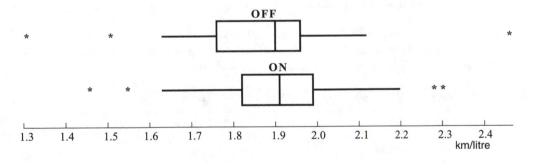

Figure 3.17 Box Plots of Aerofoil Data

3.13 Stratification

We can often be misled when we collect data in aggregate. Important effects can be disguised. Tracing information back to its sources is called *stratification*. It is the process of dividing data into subgroups, which can reveal a great deal of structure that may otherwise be unknown.

For example, a company buys raw material from two different suppliers. Stratification would involve breaking down the data into two groups, one for each supplier, and analysing them separately. A run chart for each supplier would allow a comparison of performance over time, while a stem and leaf diagram would allow their performance against some standard to be compared.

We used stratification in the aerofoil example where the data were broken down into two groups, namely aerofoil on and aerofoil off. It is clear that we should do so in this case, as this was the purpose of the exercise. It is a less obvious technique to use in many other situations.

  Export Sales

The export manager is reviewing results for the third quarter and finds that actual sales are $130,000 down on the target figure of $14,500,000. She knows that the sales figures are never exactly on target but $130,000 seems too large to ignore. She decides to investigate the source of the discrepancy, and examines the sales data for the various sales regions. The information is shown in Figure 3.18.

Region	Target	Actual	Difference
Australia	3625	3715	+90
Japan	4675	4725	+50
UK	3000	2800	-200
USA	3200	3130	-70
Total	14500	14370	-130

Figure 3.18   Sales ($000) by Region

Clearly the problem region is the U.K. Should she sack the U.K. sales manager? The export manager decides to look at the performance of the four sales representatives in the U.K. The results are given in Figure 3.19.

Sales Rep	Target	Actual	Difference
Henderson	750	780	+30
Smythe	800	550	-250
Whitaker	790	840	+50
Thomas	660	630	-30
Total	3000	2800	-200

Figure 3.19   U.K. Sales ($000) by Sales Rep

Smythe is obviously the problem. Let's get rid of him! Not satisfied, the export manager calls for information on the business that Smythe has done with each of the six outlets he is responsible for. This produces the data on Smythe's client sales given in Figure 3.20.

There are no real clues here as to why Smythe's sales are below target. Perhaps the export manager should get rid of Smythe after all. However, Smythe has been with the company for many years, and has been a good sales rep. Instead she decides to look at Smythe's sales by product line. This produces the data in Figure 3.21.

Client	Target	Actual	Difference
1	140	65	-75
2	110	70	-40
3	105	60	-45
4	130	65	-65
5	205	150	-55
6	110	140	+30
Total	800	550	-250

Figure 3.20   Smythe's Sales ($000) by Client

Product Line	Target	Actual	Difference
Export Lite	70	80	+10
Ice Chill	430	160	-270
Black Byte	250	250	0
Golden Glow	50	60	+10
Total	800	550	-250

Figure 3.21   Smythe's Sales ($000) by Product Line

Now the real cause of the problem appears to have been identified. Ice Chill does not appear to sell within the U.K. Should the product be withdrawn from the U.K. or should an attempt be made to find out why it is not selling? That is another issue, but before any action is taken it would be wise to check whether the other U.K. sales reps seem to have the same problems with Ice Chill.

Tracing information back to its sources, in the manner of this example, is an important use of the stratification principle.

 ## *Moisture Content of Coke*

The results on the percentage moisture content of coke (the fuel, not the drink!) produced at a coking plant taken daily over a 24-day period commencing on June 24 are given in Figure 3.22. The run chart for the data is given in Figure 3.23a. Important features of the data are clearly seen in this chart. For instance, the data are centred on 9.5%, and most of it lies within 0.3% of the centre. No obvious trend patterns over time can be seen. The source of the coal used in the coke production is also recorded in Figure 3.22. If this information is used to stratify the data in the run chart then a completely different picture emerges, as can be seen in the run chart in Figure 3.23b.

Day	Moisture (%)	Coal Source	Day	Moisture (%)	Coal Source
24/6	9.32	A	6/7	9.37	B
25/6	9.78	C	7/7	9.42	B
26/6	9.38	B	8/7	9.34	A
27/6	9.44	B	9/7	9.78	C
28/6	9.76	C	10/7	9.44	B
29/6	9.30	A	11/7	9.19	A
30/6	9.33	A	12/7	9.32	A
1/7	9.82	C	13/7	9.72	C
2/7	9.35	A	14/7	9.45	B
3/7	9.78	C	15/7	9.68	C
4/7	9.78	C	16/7	9.36	B
5/7	9.46	B	17/7	9.34	A

Figure 3.22　Moisture Content over a 24-Day Period

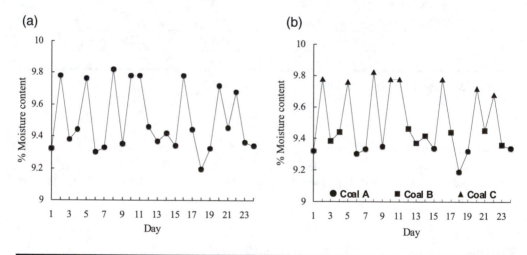

Figure 3.23　(a) Overall Run Chart; (b) Run Chart by Coal Source

Q20. *What conclusions do you come to from examining the run chart in Figure 3.23b?*

3.14 Exercises

1. A machine shop manager is studying the time taken to complete a small subassembly. Measurements, in minutes, are made at consecutive half-hour intervals. Which one of the following graphs should be used if the manager is interested in detecting skewness in the data?
   a. run chart
   b. Pareto diagram
   c. fishbone diagram
   d. stem and leaf diagram
   e. pie chart

2. In a wine tasting experiment, subjects are asked to rate the wine for sweetness using a five-point scale, ranging from very sweet to very dry. Such data are an example of
   a. quantitative data
   b. ranked data
   c. interval data
   d. attribute data

3. A personnel manager records the reasons for employees being absent from work. Such data are an example of
   a. continuous data
   b. attribute data
   c. interval data
   d. ranked data
   e. quantitative data

4. Which one of the following statements is true?
   a. box plots are used to compare sets of attribute data
   b. Pareto diagrams are used to plot data over time
   c. bar charts allow a variable to be looked at over time
   d. stem and leaf diagrams are used to plot interval data
   e. monthly sales data are an example of attribute data

5. If you are interested in studying trends over time, which one of the following plots would you use
   a. run chart
   b. stem and leaf diagram
   c. bar chart
   d. box plot
   e. Pareto diagram

6. The box plot is used primarily to
   a. look for trends over time
   b. determine problem-solving priorities
   c. compare sets of attribute data
   d. exhibit skewness in a set of data
   e. compare sets of interval data

7. The FreeAccident Transport Authority is looking into the causes of accidents on the South West motorway. They collect data on the number of accidents each week. This is an example of
   a. interval data
   b. attribute data
   c. qualitative data
   d. ranked data

Moisture	Kiln	Supplier	Moisture	Kiln	Supplier	Moisture	Kiln	Supplier	Moisture	Kiln	Supplier
14.1	1	1	21.5	1	2	23.0	1	1	26.1	1	2
17.8	2	1	14.6	2	2	13.2	2	1	17.1	2	2
16.0	1	2	14.7	1	1	18.5	1	2	18.7	1	1
10.9	2	2	15.3	2	1	11.6	2	2	19.1	2	1
21.4	1	1	17.4	1	2	17.9	1	1	21.8	1	2
16.3	2	1	12.8	2	2	14.3	2	1	16.6	2	2
20.4	1	2	24.1	1	1	19.5	1	2	19.9	1	1
12.5	2	2	14.4	2	1	14.6	2	2	11.5	2	1
18.9	1	1	23.3	1	2	22.4	1	1	17.1	1	2
14.8	2	1	18.2	2	2	15.0	2	1	14.4	2	2

Figure 3.24　Moisture Content of Logs

8. A timber processing company receives logs from two suppliers. The logs are dried in two different kilns and the company records the moisture content, in time order, of the logs coming from the two different kilns and suppliers as given in Figure 3.24.
   The moisture content data are plotted in Figure 3.25.
   a. What does the plot in Figure 3.25 tell you?
   b. Now plot the data using the information on the different kilns and suppliers. What do you now conclude from your plot?

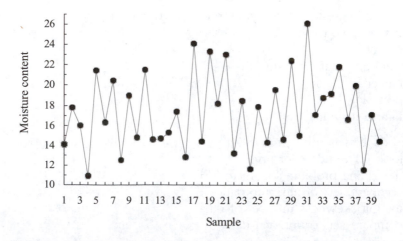

Figure 3.25　Moisture Content of Logs

9. One of the major measures of the quality of service provided by any organisation is the speed with which it responds to customer complaints. A large carpet company was receiving complaints about the installation of carpets. During a six-month period, the number of days between the receipt and resolution of a complaint is given by the data in Figure 3.26.
   a. Construct a stem and leaf diagram of these data.
   b. Write a short paragraph that summarises the service time for resolution of these complaints.
   c. What actions should be taken to further study the process of resolving complaints?

68	33	23	20	26	36	22	30
52	4	27	5	10	13	14	1
25	26	29	28	29	32	4	12
5	26	31	35	61	29		

Figure 3.26   Number of Days Taken to Resolve Complaints

10. The data in Figure 3.27 give the quality costs for a manufacturing company over the last five months of 1998.
    a. Set out these data in a table, or tables, that are easier to interpret. It may not be appropriate to present all of the data, so consider carefully which aspects of the data to present.
    b. Write a short paragraph to summarise the main features exhibited in your table(s).

11. Wilson Electronics wishes to source one of its major components from one of two suppliers. The choice has been narrowed to these two suppliers because their components were about equal in all respects and far superior to their competitors in product quality. Bill Edwards, the supply manager, decided to look at the two suppliers' performance over the past 40 weeks. Each supplier has delivered one shipment every week. The days ahead (negative value) or behind the scheduled delivery date for each of the two suppliers, LCP and FEC, are given in Figure 3.28.
    a. Construct stem and leaf diagrams of the days after schedule for each of the two suppliers. Which supplier should Wilson Electronics choose, and why?
    b. Now construct run charts of these data for each supplier, plotting both charts on the same graph. Which supplier should Wilson Electronics choose, and why?

12. The decline in the value of the British pound (£) between 1925 and 1975 is depicted in Figure 3.29.
    a. Why is this graph misleading?
    b. How would you plot these data?

13. The data in Figure 3.30 represent the number of daily calls received at an 0800 number of a large Australasian airline over a period of 20 consecutive workdays (Monday to Friday).
    Write a brief report to the marketing manager in charge of the airline's customer service concerning traffic on the 0800 number.

Prevention Costs:					
Quality audit	$270	$270	$270	$270	$270
Cost of training	$1,250	$2,300	$600	$350	$600
Improvement projects	$1,000	$2,000	$400	$400	$600
Certification	$537	$537	$537	$537	$537
Appraisal Costs:					
Inspection wages	$10,000	$10,000	$7,400	$5,000	$6,600
Audit final inspection	$3,600	$3,600	$3,200	$2,400	$2,200
Calibration costs	$250	$250	$250	$250	$250
Specification changes	$500	$300	$300	$300	$300
	Aug-98	Sept-98	Oct-98	Nov-98	Dec-98
Internal Failures:					
Running rejects	$61,100	$43,176	$96,600	$41,064	$46,526
Inplant rejects	$8,500	$25,000	$19,660	$9,700	$15,000
Rework	$2,400	$2,000	$3,600	$4,400	$2,000
Downtime	$17,480	$10,400	$7,500	$500	$9,000
External failures:					
Customer credits	$20,000	$20,000	$16,567	$12,000	$2,900
Warranty claims	$2,500	$10,000	$7,000	$0	$6,250
Totals:					
Prevention	$3,057	$5,107	$1,807	$1,557	$2,007
Appraisal	$14,350	$14,150	$11,150	$7,950	$9,350
Internal failures	$89,480	$80,576	$127,360	$55,664	$72,526
External failures	$22,500	$30,000	$23,567	$12,000	$9,150

Figure 3.27   Quality Costs

Week		1	2	3	4	5	6	7	8	9	10
Days after schedule	LCP	3.5	2.0	1.5	3.0	2.0	1.0	2.5	1.0	4.0	3.0
	FEC	0.5	1.0	0.5	1.5	1.0	0.5	1.0	0.5	1.5	0.0
Week		11	12	13	14	15	16	17	18	19	20
Days after schedule	LCP	3.5	2.0	2.5	4.0	3.0	3.5	3.5	2.5	1.5	1.5
	FEC	0.5	1.0	1.5	1.0	1.5	0.0	1.0	0.5	1.0	2.0
Week		21	22	23	24	25	26	27	28	29	30
Days after schedule	LCP	2.0	1.5	1.0	1.0	1.5	1.0	0.0	0.5	0.0	0.0
	FEC	1.0	0.5	1.5	0.5	1.0	0.5	1.5	2.0	1.5	1.0
Week		31	32	33	34	35	36	37	38	39	40
Days after schedule	LCP	1.0	0.5	0.0	0.5	-0.5	0.0	-0.5	0.0	0.5	0.0
	FEC	1.5	1.0	0.5	1.0	0.5	1.0	1.5	0.5	1.5	1.0

Figure 3.28   Delivery Times

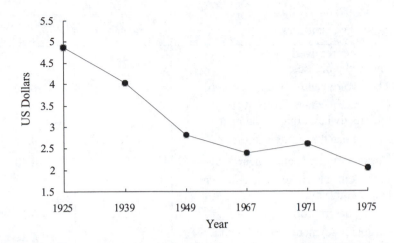

Figure 3.29   U.K.£–U.S.$ Exchange Rate, 1925–1975

Week	Day	Number of calls	Week	Day	Number of calls
1	Mon	1060	3	Mon	1603
	Tues	1370		Tues	1256
	Wed	1087		Wed	1075
	Thur	1135		Thur	1187
	Fri	1805		Fri	1060
2	Mon	1234	4	Mon	1004
	Tues	1105		Tues	985
	Wed	1168		Wed	1618
	Thur	1235		Thur	1369
	Fri	1174		Fri	1353

Figure 3.30   Customer Calls

14. Joe, the general manager of a mail order company, was concerned about the nonconformances occurring at its main customer invoicing centre. Joe decided to collect some data on the type and frequency of nonconformances, and to carry out a Pareto analysis. Over a period of one month the data in Figure 3.31 were obtained.
    a. What is the purpose of a Pareto analysis?
    b. Two Pareto diagrams could be constructed from the above data. What are they, and what are their relative merits?
    c. Construct both diagrams.
    d. Where should Joe concentrate improvement efforts?
    e. Joe is considering firing the input operator because of all the nonconformances due to input errors. Explain why this might be an inappropriate action to take.

	Type of nonconformance	Frequency	Cost per non-conformance ($)
1	Customer received goods but no invoiced processed	52	6.25
2	The invoice had no purchase number	25	0.72
3	Customer payment received but incorrectly input into the computer	23	0.85
4	Invoice had the wrong address	176	0.75
5	Customer name spelled incorrectly	202	0.40
6	Final demand sent to a customer who had already paid	8	1.05
7	Wrong description on invoice	35	0.75
8	Customer payment recorded on the wrong account	6	0.95
9	Invoice held up for more than three days in the sales department	111	1.50
10	Invoice sent to customer before the goods were	12	0.65

Figure 3.31   Nonconformance in Customer Invoices

Chapter 4

Modelling Data

4.1 Introduction

To think statistically requires an understanding of variation. As we saw in the first chapter, managers have to draw conclusions and make decisions in a variable and unpredictable world. We saw also, in Chapters 2 and 3, how data is important for effective decision making, and that it is important for data to be appropriately tabulated and displayed in order to give meaningful insights into the problem concerned. In this chapter we begin to look beyond the data to understand more about the situation that gave rise to the data. In particular, we will look at how data can be represented by simple statistical models.

A model is essentially an abstraction or conceptualisation of some situation that helps us to understand why we observe what we do. What is it that lies behind the data that we have collected to cause it to be like it is? Once we have a model that adequately represents the data we can then ask "What if?" questions. For example, suppose we collect data from a doctor's surgery on patient waiting and consultation times. Once we have postulated an appropriate model, it can be used to estimate such things as the effect of a reduction in the variation in consultation times or the provision of an extra doctor.

We will begin by looking at three different situations to motivate our consideration of how and why models of data are useful in decision-making.

🏭 *Debt Recovery*

Consider again the debt recovery problem introduced in Chapter 1, and examined in more detail in Chapter 2. A run chart of 33 months of data on the percentage of invoices paid within the due month was given in Figure 1.2. The data concerned are given in Figure 4.1, with a stem and leaf diagram in Figure 4.2.

Year	Jan	Feb	Mar	Apr	May	Jun	Jul	Aug	Sep	Oct	Nov	Dec
1990	69.0	71.0	70.5	72.0	78.0	79.0	79.5	76.0	78.0	74.0	79.0	74.5
1991	65.0	82.0	78.0	73.0	75.0	80.0	82.0	79.0	80.0	80.5	74.0	77.0
1992	69.0	76.0	79.0	77.0	76.0	80.0	83.0	80.5	79.0			

Figure 4.1   Debt Recovery Data

```
65. | 0
67. |
69. | 0 0 5
71. | 0 0
73. | 0 0 0 5
75. | 0 0 0 0
77. | 0 0 0 0 0
79. | 0 0 0 0 0 5 0 0 0 0 5 5
81. | 0 0
83. | 0
```

Figure 4.2   Stem and Leaf Diagram of Debt Recovery Data

Q1. *Why are some figures at the right end of each row in Figure 4.2 written in bold?*

Q2. *What can you see from the stem and leaf diagram in Figure 4.2 that is not easily seen in the run chart in Figure 1.2, or the data in Figure 4.1?*

Q3. *How would you describe the variability in the percentage of invoices paid within the due month?*

Julia's Diner

Julia Roberts, the owner of a local diner, is concerned at the time it takes to serve the food, once the order has been taken, during the busy lunchtime period between 12:00 and 14:00. To pacify customers who wait for a long time, she wonders whether it would be a good idea to give a $20 voucher to those customers who have waited for at least 30 minutes. She has an average of 220 lunchtime customers each week. If she goes ahead with her idea, she needs to be sure that

```
1 0
1 2 2 3 3
1 4 4 5 5 5 5
1 6 6 6 7 7 7
1 8 8 8 9 9 9 9 9
2 0 0 0 0 0 0 1 1 1 1 1 1
2 2 2 2 2 2 2 2 2 3 3 3 3 3
2 4 4 4 4 4 4 4 4 5 5 5 5 5 5 5 5
2 6 6 6 6 6 6 6 7 7 7 7 7 7 7
2 8 8 8 8 8 8 9 9 9
3 0 0 0 0 1
3 2 3 3
```

Figure 4.3   Stem and Leaf Diagram of Waiting Times

the scheme will not prove to be too expensive. Consequently, she collects data over two particular days on the time taken to serve a total of 97 lunchtime orders. The data are presented in the stem and leaf diagram in Figure 4.3.

Q4. What proportion of customers would receive discounts?

Q5. What would be the expected weekly cost of the vouchers to Julia?

Your answers to these questions are determined from the limited sample of data in Figure 4.3, and so may not reflect what is actually happening. If we can use these data to develop an appropriate *model* of the service times, then it may be possible to calculate these quantities more accurately, and so give answers to such questions as:

- Should Julia employ a part-time waiter for the lunchtime period to reduce the average waiting time? Would this strategy be cost effective?
- Should she try reorganising the kitchen operations to reduce the variability in service time? What would be the consequences of doing this?

Aerofoil Effectiveness

The stem and leaf diagram for the aerofoil data in Section 3.10, showing fuel consumption in kilometres per litre with the aerofoil fitted, is reproduced below in Figure 4.4. Generally, we are interested in whether the fitting of the aerofoil has improved fuel consumption, and what the average fuel consumption is with the aerofoil fitted.

```
1.4 | 8
  5 | 4
  5 |
  6 | 3 3
  6 | 6 6 9
  7 | 1 2
  7 | 5 5 5 6 6 7 8 9
  8 | 0 1 1 2 2 4 4 4 4
  8 | 5 5 5 5 5 5 6 6 6 8 8 8 9 9
  9 | 0 0 1 1 2 2 2 3 3 3 3 3 3 3
  9 | 5 5 5 5 5 6 6 6 7 7 9
2.0 | 0 0 1 1 2 3 3 4
  0 | 5 7 7 7 8 9
  1 | 0
  1 | 5 8 8
  2 | 0
  2 | 7
  3 | 0
```

Figure 4.4 Petrol Consumption with Aerofoil Fitted

Q6. What can you say about the pattern of variability in fuel consumption?
Q7. What is a rough estimate of the average fuel consumption for this truck?
*Q8. From the data, what do you think might be the average fuel consumption on **all** trucks of this type?*

4.2 Distributions

As you have seen in the above three examples, repeated measurements of any quantity nearly always vary. We can build up a picture of this variation as we collect data, as in Figure 4.5. This sort of picture, which is often drawn with dots rather than blocks (called a dot plot), shows us how values cluster at different points on the scale. It is rather like a stem and leaf diagram turned through 90°.

As you collect more data, the picture becomes more refined. The pattern formed is called a *distribution*. In Figure 4.6 the picture of the data is approximated

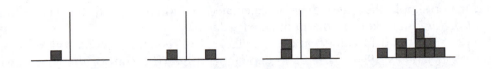

Figure 4.5   Building a Picture of Data

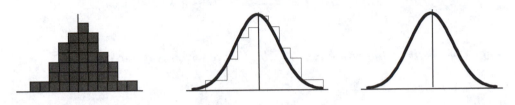

Figure 4.6   Distribution of the Data

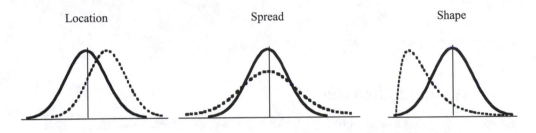

Location   Spread   Shape

Figure 4.7   Distributions Differing in Location, Spread, and Shape

by a smooth curve, which then represents the distribution of the data. The distribution curve is a *model* for the data.

Distributions can differ in the centre (or location), the amount of variation (or spread), the shape of the distribution, or any combination of these three. Changes in location, spread, and shape are depicted in Figure 4.7.

Our aim in modelling is to find a distribution that provides us with a close approximation to data arising from the process under study. This usually involves determining quantities (*parameters*) that specify different features of the distribution. These parameters usually have to be estimated from the data, using measures such as the mean and standard deviation that tell us about the *location* and *spread* of the data.

4.3   Arithmetic Mean

The run chart gives a picture of the data over time, whereas a stem and leaf diagram gives an indication of the properties of the distribution of the data. A stem and leaf diagram can show whether the distribution is fairly symmetric, or skewed; whether it has little variation or is very variable; or whether it has one, two, or more peaks.

In addition to charts and diagrams, it is also useful to summarise the data with a few well-chosen measures. The most important of these is an *average*, which measures the location of the data. The simplest type of average is the arithmetic mean, or simply the *mean*. This is calculated by adding up all the data values and dividing this sum by the number of values.

Consider the debt collection data in Figure 4.1. There are 33 data values, from 69.0% in January 1990 to 79.0% in September 1992. The sum of these values is 2525.5, so that

$$\text{Mean} = \frac{2525.5}{33} = 76.53\%$$

Q9. What is the value of the mean if the January results are excluded?

4.4   Standard Deviation

As well as defining a distribution by its location, we also need a measure to describe its spread. Consider again the debt recovery data in Figure 4.1. We can represent the data as a run chart with a horizontal line at the mean, as in Figure 4.8. Suppose that we also draw a vertical line from each data point to the mean.

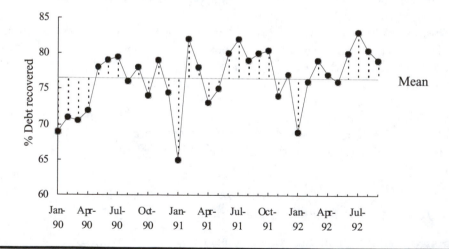

Figure 4.8   Deviations from the Mean

These deviations between the data values and the mean can be used to measure the spread or variation in the data. The greater the deviations, on average, the greater is the variation. The 33 deviations are given in Figure 4.9. Notice that some of the deviations are positive, while others are negative.

Month	1990	1991	1992	Month	1990	1991	1992
Jan	-7.53	-11.53	-7.53	Jul	+2.97	+5.47	+6.47
Feb	-5.53	+5.47	-0.53	Aug	-0.53	+2.47	+3.97
Mar	-6.03	+1.47	+2.47	Sep	+1.47	+3.47	+2.47
Apr	-4.53	-3.53	+0.47	Oct	-2.53	+3.97	
May	+1.47	-1.53	-0.53	Nov	+2.47	-2.53	
Jun	+2.47	+3.47	+3.47	Dec	-2.03	+0.47	

Figure 4.9   Deviations from the Mean

Q10. What do you think the sum of these deviations will be?

Q11. What is the reason for this?

These deviations tell us how far away each value is from the centre of the data, but if we simply average them to measure the spread of the data, we will always get zero. What we are interested in is the magnitude of the deviations, not their sign. One way of overcoming the problem with the sign is to square each deviation. Then we can take the mean of these squared deviations.

$$\text{Sum of the squared deviations} = 581.97$$

$$\text{Mean of squared deviations} = \frac{581.97}{32} = 18.19$$

Finally, if we want our measure of spread to be in the same units as our data (% in this example), we take the square root of this number. The resulting measure of spread is called the *standard deviation*. For our example, we get

$$\text{Standard deviation} = \sqrt{18.19} = 4.26\%$$

 ## Degrees of Freedom

In calculating the mean of the squared deviations, we divided the sum by 32 rather than the number of data values 33. Why do we do this? Usually we are using a set of data to model the distribution of all data from a system or process. Ideally we would like to calculate the deviations using the mean of all the available data, not just this small sample. However, the true mean is usually not available; so instead we must use an estimate of it from the data.

Because the deviations from the mean of a sample of data will always sum to zero, they are not all *independent*. For example, the sum of the first 32 deviations in Figure 4.9 is −2.47 and, since the sum of all 33 is zero, the last deviation must be +2.47. Hence, there are only 32 independent deviations. The condition that the deviations sum to zero means that we lose one degree of freedom. In general, to calculate the standard deviation from a sample of n values we divide the sum of the squared deviations by $n - 1$, the number of degrees of freedom, or independent deviations.

Later in the book we extend the idea to situations where there is more than one condition. For each condition, or parameter we have to estimate from the sample data, we lose a single degree of freedom.

4.5   Calculating the Mean and Standard Deviation

Once the mean has been determined there are three steps involved in calculating the standard deviation. They are

1. calculate the sum of the squared deviations from the mean,
2. divide by the number of degrees of freedom,
3. take the square root.

This is tedious to do manually, but many electronic calculators have built-in functions for the mean and standard deviation. However, care has to be taken to ensure that the calculation of the standard deviation involves the correct number of degrees of freedom, as some use the sample size n rather than the degrees of freedom $n - 1$.

✍ *Using Excel*

We can use Excel to calculate the mean and standard deviation. The first step is to enter the data into an Excel spreadsheet. The mean and standard deviation are then obtained from the functions *AVERAGE* and *STDEV*.

For example, suppose that the debt recovery data has been entered in cells A1 to A33. To obtain the mean, select a blank cell and enter the formula

$$= \text{AVERAGE(A1:A33)}$$

Likewise, in another blank cell, enter

$$= \text{STDEV(A1:A33)}$$

Rather than type in the function names, they can be selected from the *Insert … Function* menu. The dialogue box for *STDEV* is shown in Figure 4.10. The data in cells A1:A33 have been entered in the Number1 box, and the formula result of 4.26 is the standard deviation, as before. The dialogue box for *AVERAGE* is virtually identical.

4.6   Interpreting the Standard Deviation

What does a standard deviation of 4.26% for the debt recovery data mean? If the distribution of the data is fairly symmetric then we can say that:

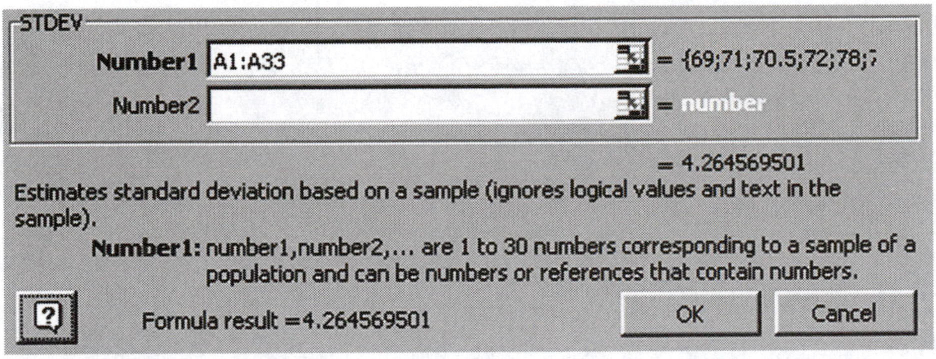

Figure 4.10   Using the STDEV Function

- approximately two thirds of the data will lie within one standard deviation of the mean,
- approximately 95% of the data will lie within two standard deviations of the mean, and
- nearly all the data will lie within three standard deviations of the mean.

For the debt recovery data we would expect, assuming nothing changes in the way invoices are processed, that about two thirds of the data will lie between 76.53 − 4.26 and 76.53 + 4.26, that is, between 72.3% and 80.8%.

Q12. *How many of the 33 data points fall within these limits?*

Q13. *What percentage of the total does this represent?*

The standard deviation is a useful measure for comparing the variation in different data sets. However, it can be misleading to compare standard deviations in sets of data with different means. For this reason, it is common to express the standard deviation as a proportion (or percentage) of the mean. This gives us the *coefficient of variation* (CV).

For the debt recovery data, the coefficient of variation is

$$CV = \frac{4.26}{76.53} = 0.056, \quad \text{i.e., } 5.6\%$$

Recall from Chapter 2 that a team was set up to work on the debt recovery problem. Data collected after the various improvement measures were put in place showed that the mean had increased to 81.6%, with a standard deviation of 3.16%.

> **Q14.** *What is the coefficient of variation after the improvement process?*
>
> **Q15.** *What effect has the improvement process had on the variability of the percentage debt recovered?*

4.7   Descriptive Statistics

As well as the mean and standard deviation, there are many other ways in which data can be described. There are other averages, such as the median that was discussed in Chapter 3, and there are other ways of measuring the spread of the data. For example, the range of the data and the interquartile range (also discussed in Chapter 3) are measures of spread, as is the mean absolute deviation (MAD).

Like the standard deviation, the MAD is calculated from the deviations. However, rather than square the deviations, we simply regard all the deviations as being positive. For example, in Figure 4.9, MAD is obtained by taking the mean of the 33 deviations ignoring the minus signs. This gives a MAD of 3.42%. MAD can be calculated very easily using the Excel function *AVEDEV*, which works in exactly the same way as *AVERAGE* and *STDEV*.

> **Q16.** *For the debt recovery data, the mean absolute deviation is smaller than the standard deviation. Do you think that this is just chance, or is there some reason why this will usually be the case?*

Many of the more frequently used measures for describing data are provided in an Excel *Data Analysis* tool called *Descriptive Statistics*. If you are unsure how to load and/or access *Data Analysis*, you should refer to Appendix A. The *Descriptive Statistics* dialogue window is shown in Figure 4.11.

Suppose the label "% Debt recovered" has been entered into cell A1, and the debt recovery data into cells A2:A34. In the *Descriptive Statistics* dialogue window in Figure 4.11, cells A1:A34 are entered as the *Input Range*, and *Labels in First Row* is checked. In our case, the data are in a single column, so we have to click on *Columns* in the *Grouped By* section. In general, it is possible to determine descriptive statistics for more than one set of data at the same time, so the *Input Range* could contain more than one column. In this case it is important to specify correctly whether the sets of data are to be read across the *Rows* or down the *Columns*.

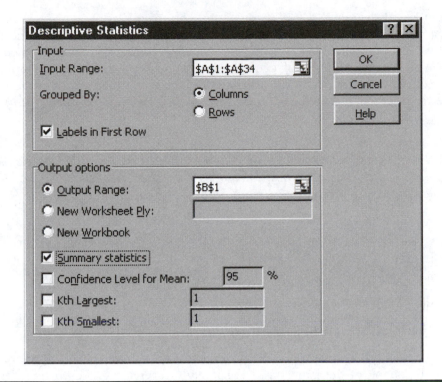

Figure 4.11   The *Descriptive Statistics* Dialogue Window

Finally, we need to specify where the results are to be written (usually an *Output Range* in the same sheet as the data), and which statistics we require. The basic option is *Summary Statistics*, which will be adequate for most purposes; so, check the box at the bottom of the window.

Using *Descriptive Statistics* for the debt collection data gives the output in Figure 4.12, some of which you will recognise, such as the mean of 76.53% and

% Debt Recovered	
Mean	76.5303
Standard Error	0.7424
Median	78
Mode	79
Standard Deviation	4.2646
Sample Variance	18.1866
Kurtosis	0.3212
Skewness	-0.8567
Range	18
Minimum	65
Maximum	83
Sum	2525.5
Count	33

Figure 4.12   Descriptive Statistics for Debt Collection Data

the standard deviation of 4.26%. Many of the other measures need no explanation, such as the minimum, maximum, sum, and count. The median and range have already been introduced in Chapter 3. The *mode* is the most frequently occurring value, and is another form of average, and the *sample variance* is simply the square of the *standard deviation*. *Skewness* and *kurtosis* are measures of the shape of the distribution, and are both beyond the scope of this book. Finally, *standard error* will be discussed in Chapter 7.

> **Q17.** *From Figure 4.12, you can see that the mean is smaller than the median, which is itself smaller than the mode. What does this tell you about the distribution of the debt collection data?*

4.8   Modelling Julia's Diner

The stem and leaf diagram of waiting times in Julia's diner, given in Figure 4.3, shows the shape of the data, but it is slightly clearer if we present the data in the form of a bar chart as in Figure 4.13. Note that we have plotted proportions along the vertical axis, given by dividing the frequency in each group by 97, the total number of meals served. In modelling the distribution of a set of data, we are looking for a smooth curve that reflects the general characteristics of the data. If we collected many more observations, the detailed pattern of the data would almost certainly change slightly; so, we want a curve that could reasonably reflect what a large sample of data might look like *in the long run*. In other words, we want a curve that represents the essential general characteristics of the data, rather than the fine detail.

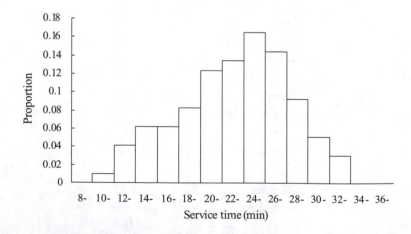

Figure 4.13   Distribution of Service Times

> ***Q18.*** *What are the general characteristics of the shape of the service time distribution in Figure 4.13?*

Suppose we superimpose a symmetric, bell-shaped curve on the service time data as shown in Figure 4.14. This curve provides a reasonably good approximation to the bar chart, and if we collected more data on service times from the diner, we might expect a bell-shaped curve to become an even better approximation of the bar chart. The curve, therefore, provides a good description or model of this situation.

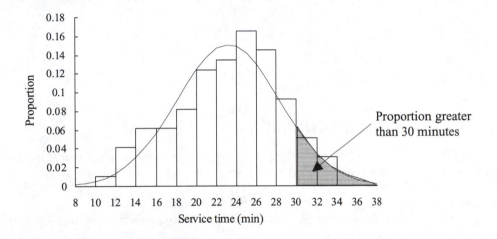

Figure 4.14   Distribution of Service Times

So, to estimate the proportion of service times that are greater than 30 minutes (the discount qualifying time), we can use the model given by this curve. This proportion, which we originally estimated by the sum of the two right-most bars on the bar chart, is now given by the area under the curve to the right of 30 minutes, as shown by the shaded area in Figure 4.14.

4.9   Distribution Shapes

The distribution of many naturally occurring sets of data is often symmetrical and roughly bell shaped. For example the distributions of the following will be approximately symmetrical and bell shaped:

- the heights of males in Australia,
- the weights of two-dollar coins,

- the time taken by an experienced machinist to manufacture a standard component,
- the scores from a large group of students on a statistics test.

However, there are other situations where the distributions will **not** be symmetric or bell shaped. For instance the distribution of:

- personal incomes,
- waiting times for hospital treatment,
- time periods between accidents in a sawmill,
- the time taken to drive from the centre to a particular suburb of a large city.

> **Q19.** *Of the three examples in Section 4.1, which do you think could most reasonably be approximated by a symmetric, bell-shaped distribution? Which least fits this pattern?*

4.10   The Normal Distribution

The most commonly used symmetric, bell-shaped distribution curve is the *normal distribution*. The normal distribution is important and useful because:

- many patterns of variation are naturally normal,
- the normal distribution is easy to work with mathematically and its properties are well known,
- in many applications conclusions drawn on the basis of the assumption of normality are not seriously influenced by small departures from the normal curve, and
- whenever a number of variable quantities is summed (or averaged), the result tends to be approximately normally distributed.

There is a complex equation that describes the normal distribution, but for our purposes the key point is that the normal curve is completely specified by the values of two quantities or *parameters*. If we know the values of these parameters, then we can find the area under the normal curve between any two limits, which represents the proportion of observations that should fall in that interval. These two parameters are

- the (arithmetic) mean, which will be denoted by the Greek letter μ (mu), and
- the standard deviation, denoted by the Greek letter σ (sigma).

For example, the mean and standard deviation of the service time data in Figure 4.3 are 22.7 and 5.1 min, respectively. The curve that we superimposed

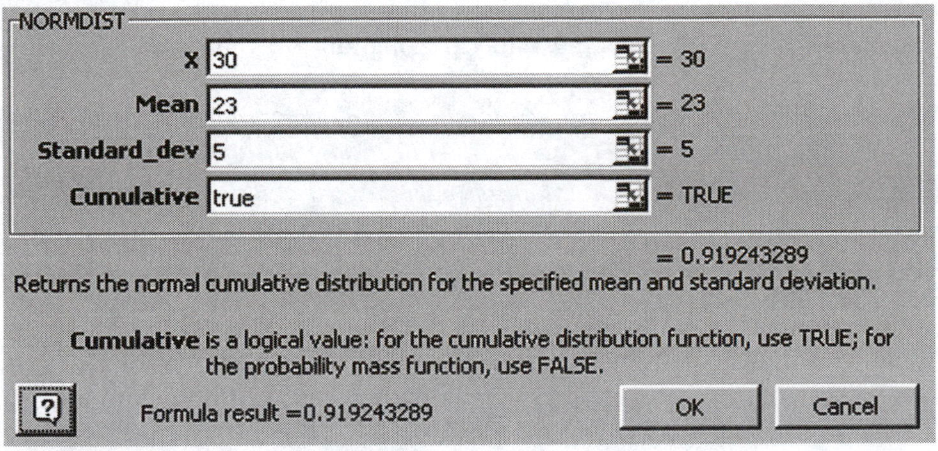

Figure 4.15 *NORMDIST* **Dialogue Box**

on the data in Figure 4.14 was in fact a normal distribution with a mean of 22.7 and a standard deviation of 5.1. So, if we can assume that the distribution of *all* service times will have a similar mean and standard deviation, a reasonable model for the service times is a normal distribution with a mean of about 23 min and a standard deviation of about 5 min.

We can use the Excel function *NORMDIST* to calculate the required areas from a normal distribution. Figure 4.15 shows the *NORMDIST* dialogue box for calculating the proportion of values *less than* $x = 30$ for a normal distribution with $\mu = 23$ and $\sigma = 5$. The fourth quantity (*Cumulative*) must be specified as *true* (or any value other than 0) or else the function will give the height of the normal curve above the horizontal axis, which is not what we want.

Figure 4.15 gives the proportion of customers who wait less than 30 min, namely about 92%. The remaining customers, about 8%, will wait for 30 or more minutes. That is, we subtract the value in Figure 4.15 from 1 to give $(1 - 0.9192) = 0.0808$, or about 8%.

> *Q20. What is your estimate now of the number of vouchers that would be issued each week? What is the expected cost of this?*

4.11 Normal Tables

Traditionally statistics has used sets of tables to evaluate areas under the normal curve, rather than an Excel function. Normal distributions come with different means and standard deviations, but we can always transform them to an equivalent *standard normal distribution* that has a mean of zero and a standard deviation of one. The standard normal distribution has been extensively tabulated, and is given in Appendix B.1. The table gives the proportion of items *greater than* any value z. Note that this is the complement of the probability given by Excel, that is, the Excel value is 1 minus the value in Appendix B.1, and vice versa.

> **Q21.** *In a standard normal distribution, what proportion of values will exceed 1? What proportion will exceed 2? What proportion will exceed 3?*
>
> **Q22.** *Look back at the interpretation of the standard deviation in Section 4.6. Can you see the basis for the statements about the proportion of values falling within 1, 2, and 3 standard deviations of the mean?*

How do we proceed if the mean is not zero, or the standard deviation is not one?

� *Standardising a Normal Distribution*

In Julia's Diner, we have a ***non***-standard normal distribution with a mean of 23 min and a standard deviation of 5 min. How can we use the table of the standard normal distribution to calculate the proportion of customers with service times greater than 30 min?

Any normal distribution can be rescaled to the standard normal distribution, as illustrated in Figure 4.16.

The procedure involves expressing any value, such as 30, as a difference from the mean in multiples of the standard deviation. This gives a *z-value* for the figure

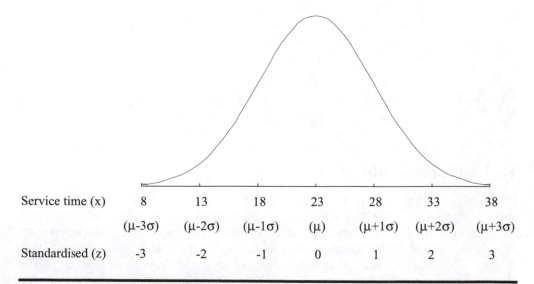

Service time (x)	8	13	18	23	28	33	38
	$(\mu-3\sigma)$	$(\mu-2\sigma)$	$(\mu-1\sigma)$	(μ)	$(\mu+1\sigma)$	$(\mu+2\sigma)$	$(\mu+3\sigma)$
Standardised (z)	-3	-2	-1	0	1	2	3

Figure 4.16   Standardising a Normal Distribution

concerned. For example, if the mean is 23 and the standard deviation is 5, then a time of 30 min is 7 min, or 1.4 standard deviations, *above* the mean. In other words, a z-value of +1.4. In general, for any value x, its z-value is given by:

$$z = \frac{x - \mu}{\sigma} \qquad \text{where } \mu \text{ is the mean, and } \sigma \text{ is the standard deviation}$$

Thus, in Julia's diner, the proportion of service times over 30 min is the same as the proportion over 1.4 in the standard normal distribution. From Appendix B.1, this is 0.0808.

Formally, the calculation would be written as follows:

$$z = \frac{30 - 23}{5} = 1.4$$

Therefore, P(Service time > 30) = P(z > 1.4) = 0.0808. That is, about 8% of customers would wait for at least 30 min for their meal to be served, as before.

 ## Two Improvement Strategies

Julia decides that she must improve service at the diner. She has two strategies in mind. One is to employ an extra lunchtime waiter, which should reduce the average service time to 15 min. The extra waiter will cost about $200 a week. Her other idea is to reorganise the kitchen operations at a one-time cost of $1000, which should reduce the variability in service time by about 40%. Which strategy would you advise her to pursue if she continues the policy of giving $20 vouchers for service that takes at least 30 min?

> **Q23.** *What would be the expected weekly saving in vouchers from the extra waiter?*
>
> **Q24.** *What would be the expected weekly saving in vouchers from the reorganised kitchen operations?*
>
> **Q25.** *Which of the two options would you advise, and why?*

A Normal Distribution Model for the Aerofoil Data

From the shape of the stem and leaf diagram in Figure 4.4, it appears that the fuel consumption data closely follows the pattern of a normal distribution. Furthermore, it is reasonable to assume that fuel consumption should vary symmetrically around some average figure, with higher and lower values becoming progressively less likely. So, a normal distribution is a plausible model that fits the data.

The mean and standard deviation of the data in Figure 4.4 are approximately 1.9 and 0.15, respectively. Based on these values, we can set up a normal model for the aerofoil data. We can then use the model to answer questions such as:

- What proportion of trips has a fuel consumption of between 1.7 and 2.2 km/litre?
- What are the quartiles of the distribution, and hence the interquartile range?
- What fuel consumption figure is exceeded on 90% of trips?

What Proportion of Trips Has a Fuel Consumption of between 1.7 and 2.2 km/litre?

We first need to determine the proportions corresponding to $x = 1.7$ and $x = 2.2$ as shown in Figure 4.17.

For $x = 1.7$,

$$z = \frac{1.7 - 1.9}{0.15} = -1.33 \quad P(x < 1.7) = P(z < -1.33) = 0.0918$$

Note that, since the distribution is symmetric, the proportion less than -1.33 is equal to the proportion greater than $+1.33$.

For $x = 2.2$,

$$z = \frac{2.2 - 1.9}{0.15} = 2.00 \quad P(x > 2.2) = P(z > 2.0) = 0.0228$$

Adding these proportions $(0.0918 + 0.0228 = 0.1146)$ gives the proportion below 1.7 and above 2.2. The proportion of trips that has a fuel consumption between 1.7 and 2.2 is therefore $1 - 0.1146 = 0.8854$, or 88.5%.

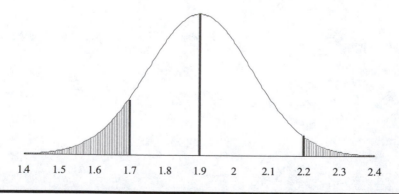

Figure 4.17   Probability that Fuel Consumption is Less than 1.7 or More than 2.2

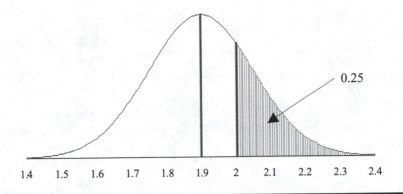

Figure 4.18 Fuel Consumption that is Exceeded 25% of the Time

What Are the Quartiles of the Distribution, and Hence the Interquartile Range?

To determine the upper quartile, we must find a value of fuel consumption (x) that is exceeded 25% of the time. From Appendix B.1, first find the value of z that gives a proportion of 0.25. Searching in the body of the table, the nearest entry to 0.25 is 0.2514 corresponding to $z = 0.67$. The next entry is 0.2483, corresponding to $z = 0.68$; so, a more accurate estimate of z is 0.675. The upper quartile is therefore 0.675 standard deviations above the mean, i.e.,

$$Q_3 = 1.9 + 0.675 \times 0.15 = 2.0 \text{ km/litre}$$

This is shown diagrammatically in Figure 4.18.

Likewise, the lower quartile is 0.675 standard deviations below the mean, i.e.,

$$Q_1 = 1.9 - 0.675 \times 0.15 = 1.8 \text{ km/litre}$$

giving an interquartile range of $2.0 - 1.8 = 0.2$ km/litre.

The quartiles of a distribution are examples of what are called *percentiles*. In particular, Q_1 is the 25th percentile and Q_3 is the 75th percentile. Percentiles for a normal distribution can be simply determined using an Excel function *NORMINV*. It gives the required value for a specified proportion (*probability*), mean and standard deviation. The *NORMINV* dialogue box for finding the lower quartile is shown in Figure 4.19; the result is 1.8 km/litre as above. Notice that the probability figure is again the proportion *below* the required value, and so the upper quartile corresponds to a probability figure of 0.75.

What Fuel Consumption Figure Is Exceeded on 90% of Trips?

This is another example of a percentile, this time the 10th percentile. This can be determined directly from Excel, with a figure of 0.1 entered as the probability in Figure 4.19. The diagram is shown in Figure 4.20.

Alternatively, using the normal table, we require first the z-value corresponding to a proportion of 0.1. From Appendix B.1, the closest proportion in the table is

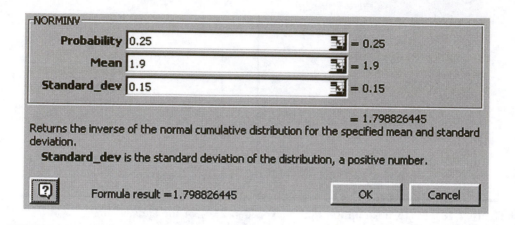

Figure 4.19   *NORMINV* **Dialogue Box**

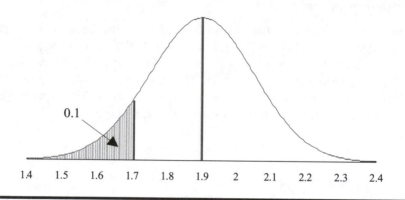

Figure 4.20   Fuel Consumption that is Exceeded 90% of the Time

0.1003 corresponding to $z = 1.28$, but this is the value that is *exceeded* 10% of the time. The value we require is therefore $z = -1.28$, that is, 1.28 standard deviations *below* the mean.

Hence, the fuel consumption figure exceeded on 90% of trips is $1.9 - 1.28 \times 0.15 = 1.71$ km/litre.

If you determine the above quantities from the data in Figure 4.4, you will find that:

- 77 of the 86 values are between 1.7 and 2.2, which is 89.5%.
- the lower quartile (the 22nd value) is 1.82 and the upper quartile (the 65th value) is 1.99. This gives an interquartile range of 0.17.
- there are 8 values (9.3%) of 1.71 or less, and there are 9 values (10.5%) of 1.72 or less. Thus the 10% point is about 1.715.

There is, therefore, fairly good agreement between the data and the above results based on a normal model, which perhaps justifies our use of a normal distribution. The results from the normal model should be more representative of the overall situation, provided, of course, that our model is the correct one.

Q26. *The manufacturers of the aerofoil have brought out a new design that is claimed to increase average fuel consumption by 10%. Assuming that their claim is correct, how would this alter your answers to the three previous questions?*

In subsequent chapters we will use various values of z corresponding to different percentages falling equally in the two *tails* of the distribution. In other words, what value of z will have a specific percentage (P) of the standard normal distribution falling between $-z$ and $+z$? A selection of these *percentage points* is given in Figure 4.21.

P	z
90%	1.645
95%	1.960
98%	2.326
99%	2.576

Figure 4.21 Percentage Points of the Normal Distribution

For instance, this table tells us that 95% of service times of customers in the diner will have z values that fall between -1.96 and 1.96, that is, times between $23 - 1.96 \times 5 = 13.2$ min and $23 + 1.96 \times 5 = 32.8$ min.

4.12 Exercises

1. For a symmetric distribution, what can we say about the mean and median?
   a. They are the same.
   b. They are possibly the same, but possibly different.
   c. They are always different.
   d. Nothing, as we have insufficient information.

2. The mean and standard deviation of a set of interval data from a symmetrical distribution are 7 and 2, respectively. Which of the following is true?
   a. Nearly all the data will be between 4 and 10.
   b. About 95% of the data lies between 6 and 8.
   c. Virtually none of the data will fall outside the interval 1 to 13.
   d. About 70% of the data lies in the interval 6 to 8.
   e. About 95% of the data will be below 8.

3. The standard deviation is a measure of
   a. location
   b. variability
   c. symmetry
   d. location and variability
   e. none of these

4. If a constant value was added to every score, the standard deviation would
   a. remain the same
   b. increase by the size of the constant
   c. increase by the square root of the constant
   d. increase by the square of the constant
   e. none of the above

5. The mean income for managers in a large organisation would be expected to be
   a. less than the median income
   b. greater than the median income
   c. the same as the median income
   d. none of the above

6. The interquartile range is a measure of
   a. location
   b. variability
   c. skewness
   d. none of these

7. From a stem and leaf diagram of 200 scores, the mean is calculated to be 70 and the median 60. This would suggest that the distribution is
   a. skewed
   b. symmetric but not normal
   c. normal
   d. none of the above

8. An Automatic Teller Machine (ATM) situated on Victoria Street dispenses cash to bank customers. The amount of money dispensed each day is approximately normally distributed with a mean of $8,500 and a standard deviation of $1,500.
   a. If $11,000 is put into the ATM at the beginning of each day, on what proportion of days will the ATM run out of money?
   b. What percentage of days will the amount dispensed be between $7,000 and $10,000?
   c. What amount of money should be placed in the ATM at the beginning of each day so that the ATM runs out of money on no more than 1% of days?

9. A large company has an annual performance appraisal system to assess its employees. Each employee is assessed on a number of different criteria related to their performance. On each criterion they are given a score on a scale from 0 to 9, where 0 means very unsatisfactory and 9 means outstanding. These individual scores are then averaged to produce an overall rating for the employee. A rating of less than 3 is regarded as "underachieving", while a rating greater than 8 is regarded as "exceptional". Underachievers and exceptional employees are the subject of special attention by senior management. In an attempt to ensure consistency, fairness, and comparability throughout the company each departmental head has to ensure that the ratings of about two thirds of the employees in his or her department are between 5 and 7. The managing director assesses departmental heads.
   a. Assuming that the distribution of performance ratings can be approximated by a normal distribution with mean 6 and standard deviation 1, what proportion of employees are expected to underachieve? What proportion will be exceptional?
   b. Comment on the performance appraisal system used by this company. What are its good and bad points, if any? Can you think of a better way of assessing performance?

10. A company manufactures a type of PVC pipe that has a minimum tensile strength specification of 52 MPa. There has been a change of supplier of raw materials, and the production manager wants to check the tensile strength of the pipes manufactured with the new materials. A sample of 50 pipes was tested and gave a sample mean of 65 MPa and a sample standard deviation of 4.8 MPa.
   a. What proportion of the pipes made using the new materials has a tensile strength below the specification limit? State any assumptions you make.
   b. With the previous supplier of the raw material the proportion of pipes with a tensile strength below 52 MPa was about 2%. Can you be confident that the change of supplier has led to an improvement?

11. Hill's Weaving Company is looking at two suppliers of yarn to be woven into men's suits. The elasticity of the yarn is an important attribute, because the yarn tends to snap if the elasticity is low. The elasticity of 40 samples of yarn has been measured from both suppliers. Supplier A's yarn has an elasticity with a mean of 28.3 and a standard deviation of 3.5, whereas supplier B's has a mean of 33.0 and a standard deviation of 5.5. If the elasticity is below 12 the yarn is unacceptable, as it will snap in use.
   a. Which supplier would you choose?
   b. Explain the reasoning for your choice.

12. A bank is looking into the amount of cash in new notes it should keep available for customer cash withdrawals. The bank always tries to pay out withdrawals in new notes. A recent study has shown that the amount of cash withdrawn by customers on any day is normally distributed with a mean of $14,000 and a standard deviation of $2,000.
   a. Suppose the bank currently starts each day with $18,000 in new notes. On what proportion of days will old notes have to be paid to customers?

b. The bank aims to ensure that used notes are issued on only 1% of days. What amount of new notes is required at the start of each day to achieve this target?

c. The bank is looking at reducing the amount of new notes it holds each day. It is examining the following two strategies:

   i. To reduce the amount of withdrawals, the bank can use various initiatives to encourage its customers to use EFTPOS (paying for goods and services at the point of sale). They estimate this will reduce mean withdrawals to $13,000, but leave the standard deviation unchanged.

   ii. Encourage customers to make withdrawals less often by higher bank charges for each withdrawal. They estimate this will leave the mean unchanged at $14,000 but reduce the standard deviation to $1,500.

If the bank wants to ensure that used notes are issued on only 1% of days, what strategy should it adopt?

13. A machine shop manager wishes to study the time it takes an assembler to complete a given small subassembly. Measurements, in minutes, are made at consecutive half-hour intervals.

a. Explain why a run chart might be an appropriate way of displaying these data.

b. What graphical method would you use if you were interested in detecting skewness in these data?

c. The mean time taken by the assembler is 12 min with a standard deviation of 2 min. Assuming the data follow a normal distribution, estimate how often the time taken would exceed 15 min.

Chapter 5

Attribute Data

5.1 Analysing Attribute Data

You will recall from Chapter 3 that attribute data simply classifies each item into one of a range of mutually exclusive categories. For example, a job meets specifications or not, a request is processed correctly or not, a family belongs to socioeconomic group A, B, C1, C2, D, or E, and so on. Attribute data is often readily available from inspection records or market surveys, and is usually quick and inexpensive to collect.

The only property that attribute data possesses is that of equivalence (or not) between any two items. For example, two families either do or do not belong to the same socioeconomic group. As a consequence, the type of analysis that can be performed on attribute data is very limited. The only average that can legitimately be computed is the mode (the most frequently occurring category). Apart from that, analysis is usually confined to examination of the pattern of occurrence between the various categories.

To gain some insight into the sort of data we are concerned with, and some of the issues that might arise, we begin by considering three examples.

🏭 *Consumer Brand Preferences*

A deodorant manufacturing company carried out a market research survey to look at customers' preference for different brands of one of their products. They took a random sample of 150 purchasers of the product and observed whether they were males or females, and which one of the three brands (A, B, or C) they bought. The results are given in Figure 5.1.

Sex	Preference			Total
	Brand A	Brand B	Brand C	
Male	13	22	28	63
Female	28	40	19	87
Total	41	62	47	150

Figure 5.1   Results of Market Research Survey

Q1. *What questions do you think would be of interest to the manufacturer of this product, and why?*

Q2. *What do you conclude from the data?*

Charnwood Casting Company

The Charnwood Casting Company has recently been exploring ways in which labour costs could be reduced, and has decided to offer an early retirement package to all employees over 60. The company has three factories in the north of England, and a sample of qualifying employees has been asked whether they are prepared to accept the package. The results, by factory, are shown in Figure 5.2.

Factory	Number of qualifying employees	Number accepting scheme
Anston	72	43
Butterfield	112	38
Carsbridge	56	27

Figure 5.2   Early Retirement Package Acceptances

Q3. *In what way are the data in Figure 5.2 different from the data in Figure 5.1 (other than context)?*

Q4. *What are the two attribute variables observed in this example?*

Q5. *What questions would be of interest, and why?*

Q6. *What do you conclude from the data?*

Arcas Appliance Company

The Arcas Appliance Company is currently considering various changes to its pension scheme. In particular, male employees' contributions are to be increased from 5 to 7% of the basic wage and in return there will be enhanced pension provision for wives and children in the event of the employee dying before reaching retirement age. It is envisaged that all new employees will be obliged to join the new scheme, but existing employees will be allowed to choose whether or not they will change from the old scheme. To assess the likely level of acceptance of the new scheme among existing employees, a survey was conducted at one of the company's factories. The results in Figure 5.3 are tabulated according to age and marital status.

Do you intend to change to the new pension scheme?	Age under 30		Age 30 or over	
	Married	Single	Married	Single
Yes	109	153	362	52
No	43	124	207	67
Don't Know	27	102	46	3

Figure 5.3 Acceptance of Pension Scheme

Q7. *What questions would be of interest, and why?*

Q8. *How are you going to deal with the "Don't Knows"? Does it matter?*

Q9. *What conclusions do you draw from the data?*

5.2   Looking at Margins

Let us start by looking at the *margins* of Figure 5.1, namely the numbers of males and females purchasing the product, and at the preferences for the three brands. These numbers are given separately in Figure 5.4.

One of the questions we could ask here is whether females are more frequent purchasers of this product than males. We see that there were 87 females and only 63 males. However, suppose we took another random sample of 150 purchasers. Would we get similar differences between females and males, or are we just as likely to get more males than females purchasing the product?

Q10. *Do you think females are more frequent purchasers of the product?*
Q11. *If they are, in what way might the company make use of this knowledge?*
Q12. *What can you say about the preferences for the three brands?*

Sex	Number
Male	63
Female	87

Preference	Number
Brand A	41
Brand B	62
Brand C	47

Figure 5.4   Sample Numbers for Sex and Brand

Sex	Expected number
Male	
Female	

Figure 5.5   Expected Numbers of Males and Females

If the purchasers of this product were equally likely to be men or women, how many males and how many females would you have *expected* your sample to include? Write down your answer in Figure 5.5.

> *Q13.* *When you compare the observed numbers in Figure 5.4 with the expected numbers in Figure 5.5, what conclusions do you draw?*

The question we have to answer is whether there is a real (*meaningful* or *significant*) difference between what we have actually observed in the sample, and what we would have expected if men and women were, in fact, equally likely to purchase the product. For example, is the observed pattern of 63 men and 87 women significantly different from an expectation of 75 men and 75 women? Think about the following two extreme situations:

- If our sample had contained 5 men and 145 women, this would surely be seen as clear evidence that women are far more frequent purchasers of this product than men.
- If there had been 73 men and 77 women, this would be close enough to the expected 75:75 for us to conclude that the difference is due entirely to chance, that is, to random variation.

So exactly where, in between these two extremes, do we begin to believe that men and women are not occurring in equal numbers? To answer this question, we need to compute a measure of how different are the actual and expected numbers. The procedure is as follows:

1. square the difference between the actual and expected number for each category,
2. divide each square by the corresponding expected number, and
3. sum the results for all categories. If this final figure is sufficiently large, the difference is *significant*.

The calculation for males and females is shown in Figure 5.6, giving a result of 3.84.

Sex	Actual number	Expected number	Calculation
Male	63	75	$(-12)^2/75 = 1.92$
Female	87	75	$(12)^2/75 = 1.92$
Total	150	150	3.84

Figure 5.6   Calculation of Chi-Square Value

Q14. *Why do we look at the squares of the differences rather than just the actual differences?*

To decide whether the final figure of 3.84 is significantly large, we need to consult a table of the (so-called) *chi-square* distribution. This will tell us the probability of the sample result (the actual numbers) having arisen purely by chance if, in this case, the purchasers were in fact equally likely to be male or female.

A table of the chi-square distribution is given in Appendix B.3. It is used as follows.

1. First identify the number of *degrees of freedom* involved in your calculation. This is similar to the number of degrees of freedom in the calculation of the standard deviation, which you will recall was given by $n - 1$. In this case, it is not one less than the number of observations, but one less than the number of categories. In our example there are two categories, male and female, so we have $2 - 1 = 1$ degree of freedom.
2. The number of degrees of freedom will identify a row of the chi-square table. Reading along this row, we see a set of increasing values, which are benchmark figures for our calculated measure (3.84). Each benchmark figure is associated with a probability, which is shown at the top of the column. So, for 1 degree of freedom, there is a 10% chance that it will exceed 2.71, a 5% chance that it will exceed 3.84, and so on.
3. Compare your calculated value to the relevant benchmark figures to reach a conclusion.

So what does it tell us in this case? Our calculated value is identical to the 5% benchmark figure. In other words, if purchasers were equally likely to be men or women, there is only a 5% chance of getting a random sample of 150 people that contains a split as unequal as 63 men and 87 women. An analogy here would be to say that there is only a 5% chance of getting as few as 63 or as many as 87 heads when a fair coin is tossed 150 times. As 5% is a relatively low probability, it must cast some doubt on the supposition that men and women are equally likely to be purchasers. Our result is said to be *significant at the 5% level*.

We can carry out a similar analysis of the preferences for brands A, B, and C. If the brands were equally popular, how many people would you have expected to purchase each brand? Write down your answers in Figure 5.7.

Preference	Expected number
Brand A	
Brand B	
Brand C	

Figure 5.7   Expected Numbers Purchasing Each Brand

Q15. *What do you conclude when you compare the results in Figures 5.4 and 5.7 subjectively? Do you think it is reasonable to suppose that, in general, the three brands are equally preferred?*

Now complete Figure 5.8 to obtain the calculated chi-square value for the brand preference results.

Preference	Actual number	Expected number	Calculation
Brand A	41		
Brand B	62		
Brand C	47		
Total	150	150	

Figure 5.8   Chi-Square Calculations for Brands

Q16. *How many degrees of freedom are there in this case?*

Q17. *What is your conclusion now about whether these three brands are equally popular?*

✍ *Using a Spreadsheet*

The *chi-square* calculations can be done very simply in Excel. The probability that the differences between what we observe and what we would expect are due to chance is given by a statistical function called *CHITEST*. This function depends only on the actual and expected numbers, which we have to enter in the spreadsheet. Note that when we use the table in Appendix B.3 we only know that the number we calculate is less than or exceeds a certain probability, or lies between two probabilities. *CHITEST* gives us the exact probability, which we can then use to judge whether the differences are real or not.

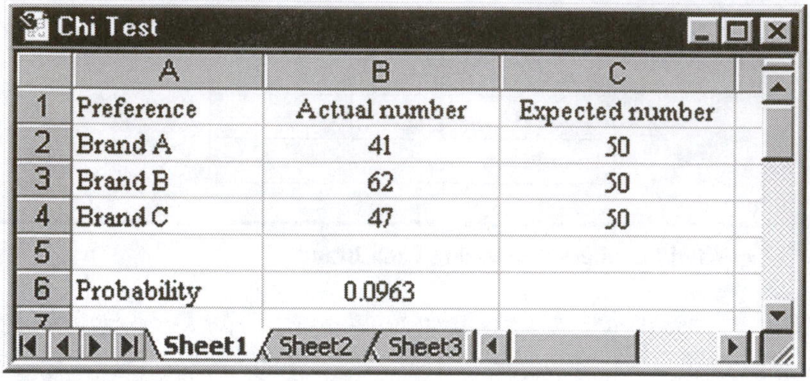

Figure 5.9   Basic Spreadsheet Data

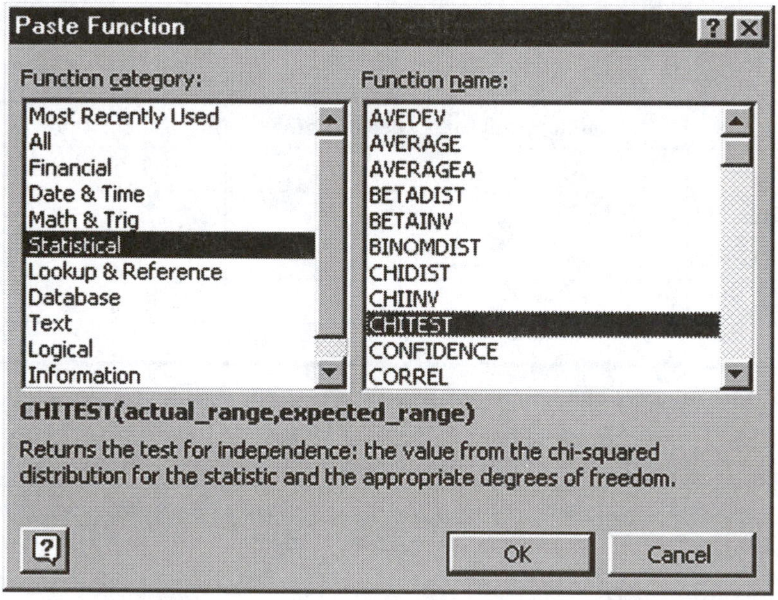

Figure 5.10   *CHITEST* Function

Figures 5.9, 5.10, and 5.11 show the relevant Excel windows for the brand preference analysis. The data are entered as shown in Figure 5.9. The labels in column A and row 1 are just for convenience; only the data in B2:C4 are strictly necessary.

The probability in cell B6 is obtained from the function *CHITEST.* With the cursor on B6, select *Insert Function* to open the *Paste Function* window as shown in Figure 5.10.

Now enter the Actual and Expected ranges in the *CHITEST* window as shown in Figure 5.11. The required probability (0.0963 in this case) is returned in B6.

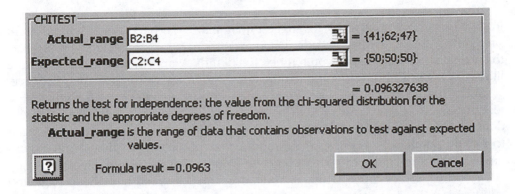

Figure 5.11 *CHITEST* Window

Q18. *How do you interpret this probability of 0.0963?*

5.3 Two-Way Tables

As well as analysing the two sets of data (sex and brand preference) independently, it is also possible to examine the preferences of males and females for each brand separately. This is likely to be just as important as our analysis in the previous section, as it will show whether or not there are different purchasing patterns between the males and females. A starting point for this analysis would be the cross-tabulation (two-way table) of sex by brand that is shown in Figure 5.1.

So far we have already concluded:

- purchasers are almost certainly not males and females in equal number, and
- there is probably not equal preference for the three brands.

Q19. *What further conclusions would you draw from the two-way table in Figure 5.1?*

Q20. *In what way might the manufacturer make use of your conclusion?*

The question that most obviously arises here is whether there is any evidence in the pattern of responses to suggest that males and females have different preferences for these three brands. There appears to be some evidence in the data that females have more of a preference for brands A and B, and males are more likely to choose C. But is this apparent pattern significant, or is it just due to chance?

We can answer this question by a straightforward extension of the chi-square test that we have just described. If the preferences for each brand were independent of sex, then we would expect each brand to have the same ratio of male to female purchasers. For instance, brand A was purchased 41 times. Since there are 63 males among the 150 purchasers, we would expect males to have purchased a fraction 63/150 of these 41 items. Similarly, females would have been expected to purchase the remaining fraction 87/150 of the items. That is,

$$\text{Expected number of males purchasing brand A} = \frac{63}{150} \times 41 = 17.22$$

and

$$\text{Expected number of females purchasing brand A} = \frac{87}{150} \times 41 = 23.78$$

Note that $17.22 + 23.78 = 41$, the total purchases of brand A. The full set of calculations for all three brands is given in Figure 5.12.

Sex	Preference			Total
	Brand A	Brand B	Brand C	
Male	17.22	26.04	19.74	63
Female	23.78	35.96	27.26	87
Total	41	62	47	150

Figure 5.12   Expected Numbers for Two-Way Table

Q21. Verify the calculations for brands B and C.
Q22. What do you conclude from an inspection of Figures 5.1 and 5.12?

Since the row and column totals in Figure 5.12 must be equal to those in Figure 5.1, we have 2 degrees of freedom for the chi-square test. For example, once we have calculated the expected numbers of males for brands A and B

(17.22 and 26.04), the remaining expected numbers are obtained by subtraction from the totals. In general the number of degrees of freedom is given by

$$\text{Degrees of freedom} = (\text{Number of row categories} - 1)$$
$$\times (\text{Number of column categories} - 1)$$

We can use the *CHITEST* function in Excel as before. The only difference is that the actual and expected numbers are now both two-dimensional arrays. This can be seen in the spreadsheet window shown in Figure 5.13.

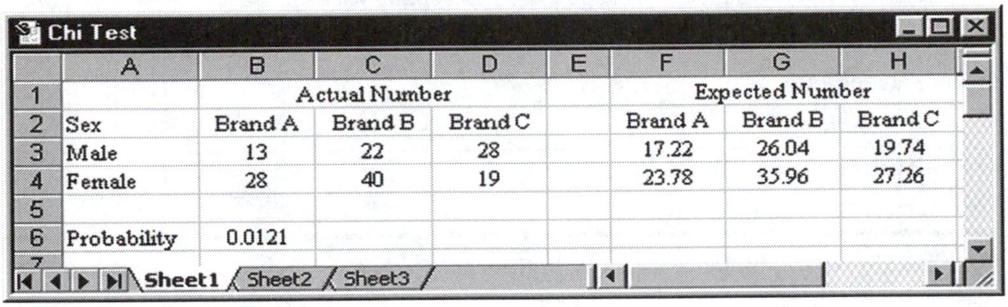

	A	B	C	D	E	F	G	H
			Chi Test					
1			Actual Number				Expected Number	
2	Sex	Brand A	Brand B	Brand C		Brand A	Brand B	Brand C
3	Male	13	22	28		17.22	26.04	19.74
4	Female	28	40	19		23.78	35.96	27.26
5								
6	Probability	0.0121						
7								

Figure 5.13 Basic Spreadsheet Data for Two-Way Table

The expected numbers in Figure 5.13 can be calculated directly from the observed numbers in cells B3:D4, as follows. The total for males is calculated in cell E3, and the formula copied into cell E4. Similarly, the Brand A total is obtained in cell B5 and the formula copied into cells C5:E5. The expected number in cell F4 is then given by the formula =$E3*B$5/$E$5, and this value is then copied into cells G3:H3 and F4:H4. Note the use of absolute cell references (see Section A.5 in Appendix A).

The chi-square probability is determined (cell B6) using the *CHITEST* function, as shown in Figure 5.14.

Q23. *How would you interpret the value of 0.0121 obtained?*

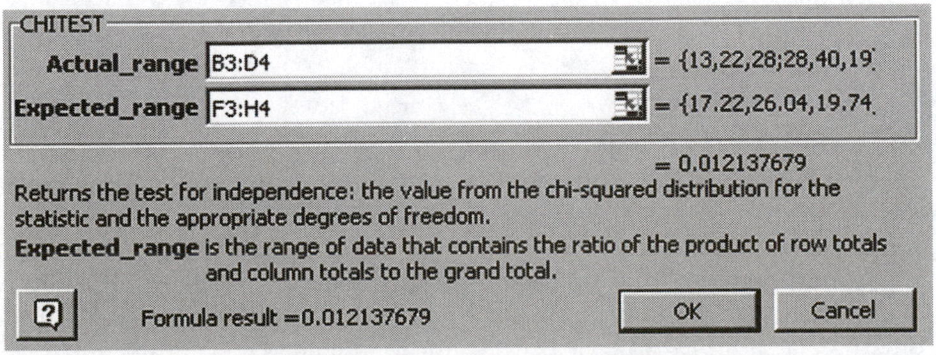

CHITEST		
Actual_range B3:D4		= {13,22,28;28,40,19
Expected_range F3:H4		= {17.22,26.04,19.74

= 0.012137679

Returns the test for independence: the value from the chi-squared distribution for the statistic and the appropriate degrees of freedom.

Expected_range is the range of data that contains the ratio of the product of row totals and column totals to the grand total.

[?]   Formula result =0.012137679   OK   Cancel

Figure 5.14 *CHITEST* Window

Charnwood Casting Company

As it stands, Figure 5.2 on the numbers of employees accepting Charnwood Casting Company's early retirement package is not a two-way table. The counts in the last column are a subset of those in the previous column. The data recast as a proper two-way table are given in Figure 5.15.

Factory	Number accepting package	Number not accepting package	Total
Anston	43	29	72
Butterfield	38	74	112
Carsbridge	27	29	56
Total	108	132	240

Figure 5.15   Two-Way Table for Early Retirement Package

Of interest in this example is whether the numbers of employees likely to accept the retirement package are the same for each factory.

> **Q24.** *Verify that the CHITEST probability to answer this question is 0.0024.*
>
> **Q25.** *What does this tell you about the employees' acceptance of the proposed early retirement package?*

5.4   Binary Data

Attribute variables with only two categories frequently arise in practice. For example, a coin is head or tail, a person is male or female, or an inspected item passes or fails a test. More generally, when we are interested in a certain key attribute, such as an item being defective, the items examined are often classified as possessing that attribute, or not. Such situations where there are only two categories give rise to *binary* data. It is usual to summarise the results of a set of binary data in terms of a measure of the frequency of occurrence of one of the categories. This measure will generally be either a *proportion* or a *count*.

For example:

- the proportion of defective items in a batch, the proportion of red beads in the paddle, or the proportion of a day's production of computer chips that do not meet specifications;
- the number of computer breakdowns in a week, the number of defective welds in 100 m of pipeline, or the number of accidents in a year. All these are *counts* rather than proportions.

So when do we use one rather than the other? In particular, it seems that we could equally well express proportions as counts. For instance, to work out the *proportion* of red beads, we clearly need to first have a *count* of how many there were in a given sample of data. In practice, we tend to use a proportion when the attribute in question occurs so many times within a defined total number of observations. With the beads experiment, knowing that the paddle contains a total of 50 beads, the occurrence of (say) 12 red beads is more informative if expressed as a proportion of 12/50 = 24%.

However, there are some situations where the total number of possible occurrences is not meaningful, or more specifically where nonoccurrence is not observable. Such situations almost always arise when the basis of observation is for a specified period of time rather than a given total number. For example, if the attribute (phenomenon) that we are recording is accidents in a factory, then how can we observe a "nonaccident?" The number of potential accident situations is extremely large (infinite?), so we cannot reduce the number actually observed to a meaningful proportion. The best we can do is to count the number of accidents occurring in a specified time period, such as a week or a month.

Just to make matters even more complicated, there are some situations in which proportions are used, but not arising from attribute data. In particular, we might be dealing with two related interval quantities that are both measured in the same terms, and where it is useful to look at one in relation to the other. For example, it is often useful to look at the monetary value of stock in relation to total turnover. This gives a ratio (or proportion), but not one that arises from counting anything. It is simply the ratio of two interval quantities, and is therefore itself an interval quantity.

5.5 Working with Proportions

In the first four chapters, the importance of understanding variation has been stressed repeatedly, and this continues to be true when working with proportions. Just as the characteristics of a sample of interval data will vary from sample to sample, so will the proportion of some attribute such as red beads or defective items. For example, when a coin is tossed 5 times, the proportion of heads obtained could be 0, 0.2, 0.4, 0.6, 0.8, or 1, corresponding to anything from 0 to 5 heads. Not all of these outcomes are equally likely, but they are all possible. If a coin is tossed fairly, we expect any sample to produce close to 50% heads and 50% tails, but there is likely to be variation around this figure from one sample to the next.

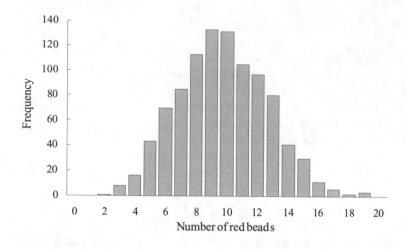

Figure 5.16   Results from 39 Beads Experiments

Consider the beads experiment, which we have conducted many times with New Zealand and British managers, and with university students. A bar chart of the results from 39 experiments, involving 975 individual results, is given in Figure 5.16. The proportion of red beads in the paddle could be anything from 0 to 1, but the evidence of Figure 5.16 seems to indicate that the proportion is unlikely to be greater than 20/50 = 0.4. From the data in Figure 5.16, the mean number of red beads per batch is 9.73, with a standard deviation of 2.90. So, we expect the proportion of red beads to be about 9.73/50 = 0.195 (i.e., about 20%), with a standard deviation of 2.9/50 = 0.058 (i.e., about 6%).

In general, we shall let p denote the *expected* proportion of items in a particular category, for example red beads in the paddle, or heads obtained when a coin is tossed a number of times. Usually p must be estimated from a set of data such as Figure 5.16, although occasionally the value of p can be assumed from the characteristics of the situation concerned. For example, it is reasonable to assume that a fair coin can be expected to show heads 50% of the time.

For simplicity let us assume we are concerned with the proportion of defective items in a batch (or sample) of a given size n (in the red beads experiment, $n = 50$). If the probability of getting a defective item remains constant, and if the out-come of one item is unaffected by (or independent of) the outcomes of any previous items, then the probability of getting a given number of defectives in the batch is given by the *binomial distribution*. This distribution is covered in detail in most books on business statistics, but we shall only be concerned with two important properties of the distribution:

1.   how the standard deviation of the proportion is influenced by p and n, and
2.   the shape of the distribution for large sample sizes.

Standard Deviation of a Proportion

We know from the theory of the binomial distribution that if the expected proportion defective is p, then the standard deviation of the proportion defective in a random sample of n items is given by

$$\text{Standard deviation} = \sqrt{\frac{p(1-p)}{n}}$$

(The standard deviation of a proportion is often called the *standard error* of a proportion in most texts.)

This means that we can infer the value of the standard deviation from an estimate of the proportion p in question. For instance, if we know that the average number of red beads is about 10, this implies a proportion of about $p = 10/50 = 0.2$. We then get an estimate of the standard deviation of 0.0566, which is in close agreement with the value 0.058 calculated from the data in Figure 5.16.

Figure 5.17 gives a plot of the standard deviation against p for the simplest case where $n = 1$. The shape of the relationship in Figure 5.17 is the same for $n > 1$; only the vertical scale changes.

> **Q26.** *Why do you think the standard deviation increases as p increases from 0, reaching a maximum at p = 0.5, then decreases until it is equal to 0 at p = 1? Is this intuitively reasonable? (It might help to consider what happens if the proportion of red beads in the batch of raw material is close to 0 or 1.)*

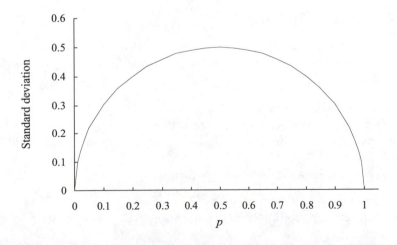

Figure 5.17   Standard Deviation of a Proportion

✥ *Normal Approximation for Proportions*

From the shape of the bar chart in Figure 5.16, it seems that the distribution of the number of red beads in the paddle follows quite closely the pattern of the normal distribution described in Chapter 4. This would still be true if the data were expressed as proportions; for example, 10 red beads becomes a proportion of 10/50 = 0.2. Working with proportions simply changes the horizontal scale by a factor of 50, the total number of beads in the paddle. So, we can say that the proportion of red beads is approximately normally distributed with a mean of about 0.2 and a standard deviation of about 0.06. The normal distribution gives a good description, or *model*, for the outcome of the beads experiment. We could, therefore, use this model to estimate the likelihood of getting, say, more than 15 red beads, that is, a proportion greater than 0.3.

But is it just coincidence that the normal distribution gives a good approximation to the beads data, or is there a more general process at work here?

It can be shown that the normal distribution gives a good approximation to any binomial distribution provided the sample size n is large. As you can see from the beads data, it is very good with $n = 50$. There are some other rules of thumb that should be considered before using this normal approximation, which are related to p, the proportion of defectives. The normal approximation is very good, even for small n, when p is close to 0.5. You should exercise caution, however, when p is very small or very large, for then the normal does not provide a good approximation.

To further illustrate the normal approximation, let us take samples from a situation in which the probability of getting a defective item is $p = 0.5$. As an analogy, we can think of an experiment where we toss a coin a number of times and we are interested in the proportion of heads. The bar charts for samples of sizes $n = 1, 2, 4$, and 8 are given in Figure 5.18.

You can see that, even with only a sample of size $n = 4$, the "bell-shaped" pattern of a normal distribution is beginning to emerge. With a sample of size $n = 8$, the approximation to the normal distribution has already become quite good.

5.6   Working with Counts

Recall from Section 5.4 that count data usually arise when we are interested in the number of occurrences of some event during a specified period of time, such as the number of

- motorists caught by a speed camera,
- passenger complaints received by British Airways,
- people entering the accident and emergency department at Waikato hospital,
- telephone calls arriving at the switchboard at Loughborough University,
- cars arriving at the tollbooths at Sydney harbour bridge.

In all these examples, the count in question will relate to some appropriate period of time.

Assuming that conditions stay the same over the period in question, in all of these situations the probability of getting a certain number of occurrences is given

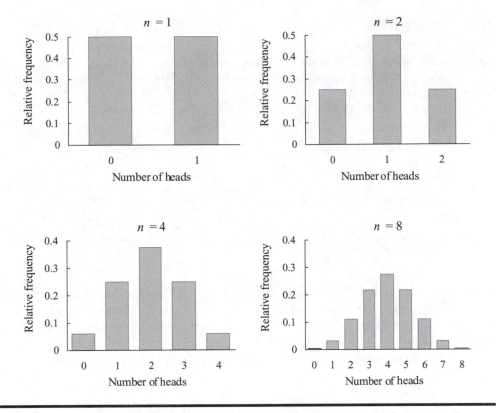

Figure 5.18 Bar Charts for Samples of Size 1, 2, 4, and 8

by the *Poisson distribution*. Details of this distribution can be found in most statistics books but, as with the binomial distribution, we shall only be concerned with:

1. the relationship between its mean and standard deviation, and
2. a normal approximation for this distribution.

Standard Deviation of a Count

Suppose we are concerned with a count of some quantity such as the number of insurance claims per year, defects per day, or accidents per quarter. From the theory of the Poisson distribution the standard deviation (or *standard error*) of this count is given by $\sqrt{\mu}$, where μ is the average rate of occurrence. For example, if the daily average number of defects is 25, then the standard deviation of the number of defects per day is 5. So, as the average count increases, so does the variability as measured by the standard deviation.

Is this relationship between the average rate and the standard deviation intuitively reasonable? Consider the occurrence of accidents in two different situations. In a small workshop, the average number of accidents is likely to be very low, say about 1 per month, while in a large timber company the average might be 30 per month.

Q27. *How many accidents each month might we typically get in these two situations?*

Q28. *Would you conclude that when the accident rate is low we get little variability in our results, while a higher accident rate leads to greater variability?*

For a second example, consider the abundance of oysters in the seas around New Zealand. In some areas there will be no oysters, so zero on average with no variability. In other areas oysters will be rare, with very low numbers on average and, hence, little variability in the numbers found. Around Bluff, oysters are found in huge quantities, and the numbers to be found will vary enormously from day to day.

Normal Approximation for Counts

Suppose that we have collected data over a 6-month period (182 days) on the number of claims per day from an insurance company, giving the bar chart in Figure 5.19. As with proportion data, this is similar in shape to a normal distribution as can be seen from the superimposed normal curve. The fit here seems less good than in Figure 5.16, but it must be remembered that there were 975 individual results in Figure 5.16, whereas here there are only 182. If more data were available,

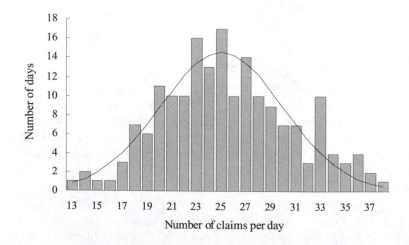

Figure 5.19   Normal Curve on Daily Number of Claims

it is likely that some of the "irregularities" in Figure 5.19 would disappear and the normal distribution would give a better model for the data.

From the data, the average number of claims is 25.5, with a standard deviation of 5.3, which conforms well to the Poisson distribution model (the standard deviation is approximately equal to the square root of the mean). However, as with the earlier beads data, it is unlikely that this information would be available in practice, but we can estimate the standard deviation to be about 5 based on an assumption that the mean is about 25. Using the normal model, we can now estimate the probability of (say) more than 30 claims per day from the area under the normal curve above 30.5. This gives

$$P(> 30 \text{ claims}) = P\left(z > \frac{30.5 - 25}{5}\right) = P(z > 1.1) = 0.136$$

Thus, there should be more than 30 claims on about 14% of days. There were actually 34 days when the number of claims exceeded 30, a proportion of 18.7%. In this case, the model underestimates the actual proportion, as can be seen from the right-hand tail of Figure 5.19, where all but one of the values above 30 had a frequency greater than that predicted by the normal model. Had more data been available, it is likely that the actual proportion of days with more than 30 claims would be closer to the predicted value.

In general, a normal distribution should provide a good model for count data as long as the average rate of occurrence is reasonably large. In most cases an average greater than 10 should be sufficient.

Q29. *There were 21 days out of 182 (i.e., 11.5%) with less than 20 claims. From Figure 5.19, would you expect the normal model to overpredict or underpredict this proportion? Why?*

Q30. *Check that you are right by estimating the proportion of days with fewer than 20 claims based on a normal model with a mean of 25.*

These relationships regarding the standard deviation of a proportion or a count will be exploited in Chapter 7 on margins of error for proportions, and in Chapter 12 on control charts for attribute data.

5.7 Exercises

1. The variability of a proportion *always* decreases as
   a. the sample size increases
   b. the sample size decreases
   c. the value of the proportion increases
   d. the value of the proportion decreases
   e. both (a) and (c) above
   f. none of these

2. Data are collected on accidents in a factory. The number of accidents could be recorded on a weekly or monthly basis. Compared with the number of accidents per week, the variability (standard deviation) of the number of accidents per month will be
   a. the same
   b. a quarter
   c. a half
   d. twice
   e. four times

3. A consumer magazine is interested in determining the relationship between the size and make of car. From a random sample of 1000 recent buyers of cars, the magazine classifies each purchase with respect to size and the manufacturer of the purchased car. An analysis of the results shows that the two classifications, size and manufacturer, are related. This means that if we know which size car a purchaser chooses, then we
   a. know that some manufacturers do not sell small cars
   b. have a clue about which manufacturer the buyer will choose
   c. will have information about whether the buyer is male or female
   d. still have no idea about which manufacturer the buyer will choose

4. It has been suggested that the public's confidence in big business is closely tied to the economic climate. When business is contracting and unemployment is increasing, public confidence is low. When the opposite occurs, public confidence is high. In a market research study, a random sample of 810 people in employment were asked questions about their job security and their degree of confidence in big business. The results are given in Figure 5.20.
   a. Explain carefully what you regard as the most important question that should be asked.
   b. The corresponding expected numbers are given in Figure 5.21. Explain precisely what the data in this table indicate. Verify that the value in the first cell is 102.0.

Confidence in Big Business	Job security			
	Very secure	Moderately secure	Slightly insecure	Very insecure
A great deal	120	53	10	6
Only some	255	168	41	22
Hardly any	62	47	15	11

Figure 5.20   Observed Job Security and Business Confidence

Confidence in Big Business	Job security			
	Very secure	Moderately secure	Slightly insecure	Very insecure
A great deal	102.0	62.5	15.4	9.1
Only some	262.2	160.8	39.5	23.4
Hardly any	72.8	44.7	11.0	6.5

Figure 5.21   Expected Job Security and Business Confidence

    c. From the above two tables, a chi-square value of 14.55 is calculated. How many degrees of freedom are associated with this value?

    d. What conclusions do you draw from this analysis? Relate your conclusions back to the suggestion made in the first sentence of the question.

5. Fred Smith, the personnel manager of the Wellington Steel Company, is concerned about the level of absenteeism among the company's weekly paid workers. He chooses 3 representative weeks over the last 12 months, and records the number of absentees for each day of the week, producing the data in Figure 5.22.

    a. What is the pattern that seems to be suggested by the data?

    b. Is there any evidence in the data of differences in absenteeism rates on different days of the week?

    c. What should Fred do about this problem?

Monday	Tuesday	Wednesday	Thursday	Friday
34	24	22	24	36

Figure 5.22   Weekly Absenteeism

6. The London Tool Company employs three accounts clerks, Albert, Bernard, and Colin. Recent concerns about the number of incorrectly prepared invoices have caused the accounts manager to examine the clerks' accuracy. Over the last 2 weeks, the number of invoices prepared by each clerk, and the number in error, are given in Figure 5.23.

    a. What is the overall error rate, that is, the proportion of invoices containing errors?

    b. If the clerks had an equal "error rate," how many invoice errors would you expect for each clerk?

	Albert	Bernard	Colin	Total
Invoices prepared	164	219	217	600
Errors involved	13	21	14	48

Figure 5.23   Errors in Invoices

   c. Is the difference between what occurred and what you expect on the basis of an equal error rate significant?

7. In Section 5.1 we discussed the changes the Arcas Appliance Company is currently making to its pension scheme. The data are reproduced in Figure 5.24.
   a. Assuming that these results may be regarded as a random sample of all the company's male employees, examine whether there is evidence of a different attitude to the new scheme depending on the
     i. age of the employee (ignoring marital status); and
     ii. marital status of the employee (ignoring age).
   b. Are these results consistent with what you would have anticipated?
   c. Think about how you have dealt with the "Don't Knows." Are there any other ways of dealing with them?

Do you intend to change to the new pension scheme?	Age under 30		Age 30 or over	
	Married	Single	Married	Single
Yes	109	153	362	52
No	43	124	207	67
Don't Know	27	102	46	3

Figure 5.24   Acceptance of Pension Scheme

8. In a market research survey concerned with home loans, a random sample of 262 people who had taken out a first mortgage in the last year was interviewed. From the interviews, the data in Figure 5.25 were collected.
   a. Does there appear to be any difference between the size "profile" of loans made by banks and building societies?
   b. Use the chi-square test to examine this formally.
   c. Does it make a difference to your conclusions if the comparison were made simply between loans of less than or equal to £50,000 and those more than £50,000?
   d. Write a short paragraph describing your findings.

Source of loan	Size of loan		
	< £20,000	£20,000 - £50,000	> £50,000
Bank	10	14	25
Building Society	54	83	76

Figure 5.25   First Mortgages

9. In the Beckan Microchip Company, the average percentage defective expected from a process that produces Extel 967 microchips has been found to be around 3%. They produce about 10,000 of these chips per week. How often will the weekly percentage defective exceed 3.5%?

10. The Statewide Insurance Company is monitoring the number of insurance claims per quarter. Suppose that over a period of a number of years the average number of claims per quarter was 7000. What would be the probability that in the next quarter there would be over 7200 claims?

11. In question 4 in Chapter 1, you were asked whether, in the beads experiment, the greatest variability in the results came from having 5%, 10%, 20%, 40%, or 80% red beads in the raw material.

a. Justify your answer by considering the standard deviations of the different proportions of red beads.

b. Compare your answer with the one you gave in Chapter 1.

Chapter 6

Sampling

6.1 Collecting Data

We have looked at the different types of data, and the reasons why we collect data, in Chapter 3. In this chapter we consider how to collect data. Most organisations collect a variety of data that is relevant to their operations. Companies are likely to collect sales data, customer data, and data about their employees. Universities collect data about students and their performance. National governments collect data about a vast array of things, such as age distribution, the country's trading and business activity, and its economic performance.

Should New Zealand Become a Republic?

In August 2000, the New Zealand Herald newspaper invited its readers to participate in a poll about whether New Zealand should become a republic. Readers were asked to call an 0900 number, at a cost of about $1 a minute, and register their opinion. After the poll, they published the results given in Figure 6.1.

Figure 6.1 Republic of New Zealand?

Q1. *Do you regard this as a representative sample of the New Zealand population? Why, or why not?*

Q2. *Is it reasonable to conclude that the majority of the New Zealand population are against a republic?*

Population Census

Many countries try to measure the size of their population on a regular basis by conducting a population census, usually every 10 years. For example, the U.S., U.K., and New Zealand will each conduct a population census in either 2000 or 2001.

Q3. *Why is a population census important? Are there any disadvantages of such a census?*

Q4. *Is there any other way of reliably obtaining the same information?*

Superannuation Referendum

In 1997 a referendum was carried out in New Zealand on the issue of compulsory superannuation. A postal ballot of all people on the Electoral Register was carried out. They were asked whether they were for or against a compulsory superannuation scheme.

Q5. *What are the disadvantages of such a referendum?*

Q6. *Can you think of an alternative method of collecting the information? What are the advantages and disadvantages of your method?*

⚒ *Market Research Survey*

A sculpture exhibition was held in an enclosed area in Battersea Park in London. The organisers wanted to sample people attending the exhibition in order to determine who was attending, where they came from, and how they found out about the exhibition. This information will be useful in planning and marketing future exhibitions.

> **Q7.** *Write down two ways in which such a sample might be taken. What are the relative merits, and disadvantages, of your methods?*

6.2 Sampling

In the superannuation and market research examples, the data required were, or could have been, collected by taking a selection from the population concerned. Any selection or subset of some population is called a *sample*. Sampling is very often the only practical way of obtaining data about large populations in a cost effective and timely way.

Suppose in the superannuation referendum we decided to estimate the proportion favouring some form of compulsory superannuation by taking a sample of eligible voters.

> **Q8.** *What factors do you think will influence the choice of the size of the sample?*
>
> **Q9.** *How would you choose a sample of, say, 5000 voters?*
>
> **Q10.** *What are the advantages and disadvantages of sampling in this case?*

6.3 Sampling Concepts

Sampling is used to get information about a population or process. The reasons we use a sample are because it is often too expensive, time consuming, practically impossible, or simply unnecessary to canvass the whole population. For example:

- A census or referendum is expensive. The required information can usually be obtained from a sample. Further, it is often more accurate to take samples, since the resources available can be focused on the smaller sample rather than across the whole population.
- It is time consuming and costly to inspect every invoice raised, or widget produced.
- Quick results may be needed on an important issue, such as a decline in sales figures.
- It is impossible to study the whole population if the sampling method is destructive, for example, when cutting fruit in half to measure fruit quality, or puncturing a can of baked beans to measure the bacterial count, or when the lifetime of an electric light bulb is required.

In order to facilitate a discussion of sampling methods we need to define some terms.

A ***population*** is a set of all the individuals or items in a group of interest.

- This may be a *finite* collection, such as the population of students enrolled in a particular course.
- It may be an *infinite* collection, or a very large group, such as the population of one-dollar coins in circulation.

A ***sample*** is a subset of the population. For example, a subset of the students on a course that form a tutorial group, the voters that live in a certain area of a parliamentary constituency, or the first hour's production of extruded items in a plastics factory are all samples of some particular population.

A ***parameter*** is a measurable characteristic of the *population*, for example, the population mean or the population standard deviation.

A ***statistic*** is a measurable characteristic of the *sample*, for example, the sample mean or the sample standard deviation.

In principle any subset of the population is a sample, no matter how it is selected. In practice, however, we need to select the sample carefully so that it is representative of the population from which it is taken. If a sample is properly selected, particular population characteristics should be reflected in the corresponding sample characteristics. In contrast, a sample that is not properly selected is likely to be nonrepresentative and give a distorted picture of the population.

The aim of sampling is to *choose* a subset of the population from which we *calculate* appropriate sample statistics that accurately *estimate* population parameters. This *choose–calculate–estimate* process is depicted in Figure 6.2.

In Chapter 4 we saw how to calculate various statistics, such as the mean and standard deviation, from a sample of data. In the next chapter we will consider some of the issues involved in estimating population parameters. In the following

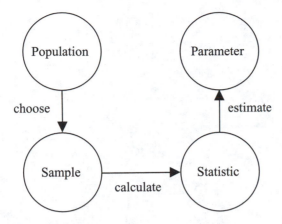

Figure 6.2 Sampling Definitions

sections of this chapter we will look at different ways of choosing an appropriate sample of data.

6.4   Choosing the Sample

There are a number of ways in which we can select a sample from a population. In the following experiment we consider some of the issues involved in the choice of a sample.

 Wood Experiment

This experiment involves estimating the mean weight of a population of 100 blocks of wood of varying shapes and sizes, but of the same thickness. The blocks range in weight from a minimum of 5 g to a maximum of 105 g, with weights recorded to the nearest 5 g. A pictorial representation of these 100 blocks is given in Figure 6.3.

We could, of course, calculate the mean of all 100 blocks to determine the *population* mean. However, for the purpose of this experiment, we will suppose that the population mean is unknown and we wish to estimate it from a sample of 10 blocks.

We shall use two different methods of sampling, namely:

- a judgment sample, and
- a random sample.

We will describe various procedures for both random and judgmental sampling in more detail in Sections 6.6 and 6.7. For the present, we will say simply that a *judgment sample* is one where a person uses their best judgment to select a sample they consider to be representative of the population being sampled. In contrast a *random sample* is chosen completely objectively without any reference to the population being sampled.

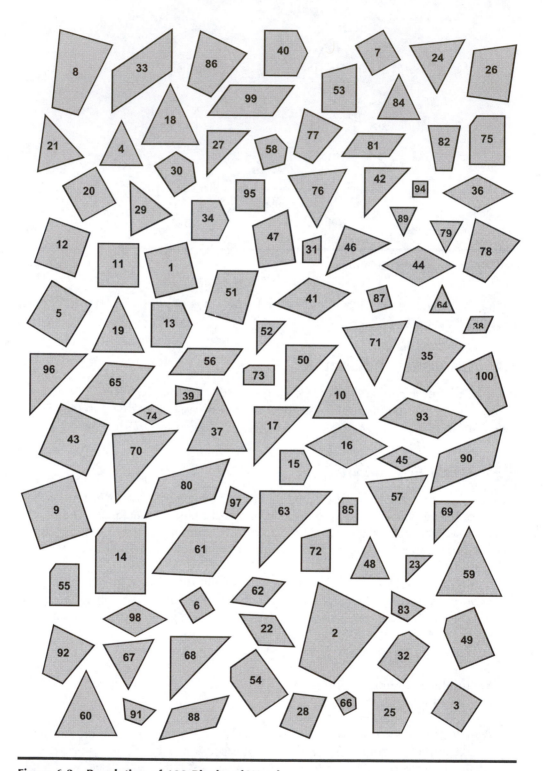

Figure 6.3   Population of 100 Blocks of Wood

> **Q11.** *What is your instinctive assessment of these two sampling methods? In particular, which method will be more accurate in estimating the mean weight of the population?*

 ## Judgment Sample

The objective is to get an accurate estimate of the mean weight of the population by taking a judgment sample. Look carefully at the 100 shapes in Figure 6.3.

1. Choose a sample of 10 of these that you regard as representative of the population. Write their numbers in the *Block Number* column in Figure 6.4.
2. The weights of all 100 blocks are given in Figure 6.8 in Section 6.9. Using this data, write the *Weight* of your chosen blocks in Figure 6.4.
3. Calculate the mean weight.

	Block number	Weight
1		
2		
3		
4		
5		
6		
7		
8		
9		
10		
Mean weight		

Figure 6.4   Judgment Sample

 ## Random Sample

Now select a second sample using a random approach. To do this use the random number table in Appendix B.4 as follows:

1. Choose a starting point in Table B.4 corresponding to your birthday. The day gives the row and the month the column. For example, if your birthday is 27 March then start with the number in the 27th row and 3rd column; that is 67. Write this number and the next 9 numbers working down the column (go to the top of the next column if necessary) in the *Block Number* column in Figure 6.5.
2. Again using the data in Figure 6.8, write the *Weight* of your chosen blocks in Figure 6.5.
3. Calculate the mean weight.

	Block number	Weight
1		
2		
3		
4		
5		
6		
7		
8		
9		
10		
Mean weight		

Figure 6.5   Random Sample

Q12. *The population mean weight of all 100 blocks is 39.4 g. Which of your estimates is closer to this population mean?*

If you are able to take part in the wood experiment with a group of students, construct a stem and leaf diagram of the results obtained by everyone in the group using the template in Figure 6.6. Notice that there are two rows for each stem. The first row is for the low units (0–4) and the second row for high units (5–9).

The wood experiment has been carried out with a class of 243 students. These students actually inspected a population of real blocks of wood, whose weights were the same as the values in Figure 6.8. Each student selected two samples exactly as you have done, and calculated the mean weight of each sample. The results of the experiment are given in the bar chart in Figure 6.7.

Using either your stem and leaf diagram in Figure 6.6 or the bar chart in Figure 6.7, answer the five questions below.

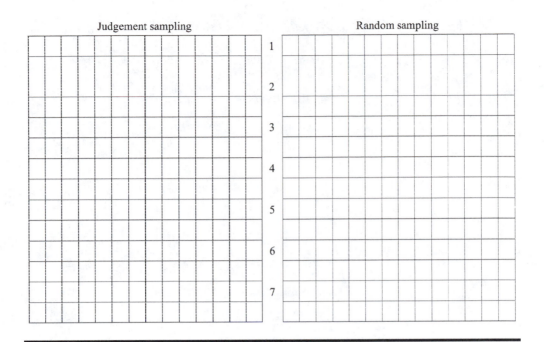

Figure 6.6   Stem and Leaf Diagrams of Judgment and Random Sample Means

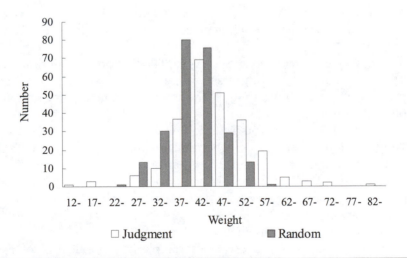

Figure 6.7   Means from 243 Judgment and Random Samples

Q13. *Roughly estimate the mean of each distribution.*

Q14. *What can you say about the amount of variation in the results from your two sets of samples?*

Q15. *Which sampling method is better in estimating the population mean, and why?*

Q16. *In Figure 6.7 the judgment sample tends to overestimate the population mean. Can you think of a reason why this is the case?*

Q17. *What are your overall conclusions from this experiment?*

6.5   Sampling in Practice

Whenever we collect data we usually have to sample the process under study. Sampling is used in most, if not all, activities. Some examples are

- *monitoring industrial processes*. As an illustration, imagine an injection-moulding machine making plastic pipes. In order to check whether the machine is producing on target, the internal diameter of a sample of five pipes is obtained every hour.

- *auditing invoices*. Auditors sample invoices to check the level of errors. They may aim to check the higher dollar-valued invoices more thoroughly, where there is more scope for large errors.

- *product design*. A company is about to develop a new style of hairbrush. It conducts market research to find out which brands the consumers are already aware of, and which aspects of its appearance and performance are most important to them.

- *political opinion polls*. These polls sample the population of registered electors who are regularly questioned to obtain their preferences for political parties.

- *environmental monitoring*. Waste is treated before being discharged into a stream. Regular samples of stream water are taken to check the level of contaminants. As fines are imposed for high levels of contaminants, it is important that the level of contaminants does not exceed the legal limit.

- *market research*. Most organisations consider it important to have detailed knowledge of their customer base. They frequently conduct surveys to find out why customers bought their products or services. This is often accomplished by using representative samples of customers to form customer panels and focus groups.

6.6   Random Sampling

A random sample is one that is selected objectively by some random process, where there is no element of judgment on the part of the sampler over which particular items are chosen from the population. Random samples are sometimes referred to as *probability* samples because the method of selection used implies that the probability of an individual item, or a group of items, being selected is known.

Q18. *Consider the example at the beginning of this chapter concerning whether New Zealand should become a republic. Is the sample of responses a random one? Why, or why not?*

Q19. *Is the sample of responses obtained in the superannuation referendum a random one? Why, or why not?*

There are a number of ways in which random samples can be obtained.

Simple Random Sampling

The random sample in the wood experiment is an example of a simple random sample. In such a sample, each possible sample of a given size is equally likely, or has the same chance as any other, of being selected. This is one of the most common probability-based sampling schemes. In order to take such a sample we must:

- Have a list of the members of the population; this is called a *sampling frame*. For example, the electoral roll is a sampling frame for the voters in an electorate.
- Decide on the sample size. This will be influenced by the level of accuracy needed. This is discussed in detail in Chapter 7.
- Choose items from the sampling frame using some random process. For example, using a table of random numbers such as the one given in

Appendix B.4. Each group of items of the chosen size then has the same chance of being selected for the sample.

Stratified Sampling

The population is first divided into natural groups or *strata*. Then a simple random sample is taken from each stratum. The number sampled from each stratum is often chosen to be proportional to the total number in that stratum. Additionally, it should take into account the variability in each stratum. For instance, if the members of a stratum are relatively homogeneous then only a small sample is required. A more variable stratum, however, requires a larger sample to give an accurate representation of the members of that stratum. Cost is another possible consideration in determining the sample size in each stratum. Stratified sampling gives more precise estimates of population parameters than a simple random sample of the same size when the strata are different and we know something about the composition of each stratum, such as the size and variability.

Suppose we wish to determine the average amount of money borrowed by undergraduate students at a university. The university has 15,000 students, 60% of whom are females spread equally throughout the years of study. In this particular year, 40% of all students are in the first year, 30% in the second, and 30% in the third year. Suppose you want to survey a sample of 120 students.

Q20. Why might we take a stratified sample? Why not a simple random sample?

Q21. How would you take a stratified sample of 120 students?

Systematic Sampling

Suppose we have a list of the items in the sampling frame. With systematic sampling we choose items at regular intervals from the list, such as every 5th or 8th item, with the starting number chosen using random numbers. For example, if our sampling frame is a telephone directory, then we might choose every 80th entry in the book, starting at the top of the second column on page 43. If the items in the sampling frame are listed in random order, systematic sampling is as good as simple random sampling, but is generally much easier to administer than simple random sampling. Once you have located the starting point, the procedure is straightforward.

A potential problem with this type of sampling is *periodicity*. That is, if there is a cycle in the data we can get badly biased estimates, especially if the cycle has the same periodicity as the sampling interval. For example, suppose we take a sample every Wednesday to estimate yearly sales at a retail outlet. If the pattern of sales differs from one day of the week to another, then our estimates could be seriously biased.

> **Q22.** *Returning to our example on sampling students at a university to determine the amount of money borrowed, how would you take a systematic sample of 120 students?*

✍ *Cluster Sampling*

This form of sampling can be used when the population falls naturally into various groups or *clusters*. For example, the clusters may be geographical areas, or students on different courses or days of the week. To obtain a sample from the particular population, some of the clusters are selected and a random sample, or census, is then taken from each of the selected clusters.

> **Q23.** *What are the advantages of cluster sampling?*
>
> **Q24.** *What are the potential disadvantages of cluster sampling? In what circumstances could cluster sampling give inaccurate results?*

Considering the four methods of random sampling described above, suggest an appropriate sampling scheme for the following two situations.

> **Q25.** *How would you take a random sample of 1000 people from an electoral roll in order to estimate the support for different political parties?*

> **Q26.** *The University of Loughborough wishes to interview a random sample of its students to find out why they chose to study at the university. How might such a sample be chosen?*

6.7   Nonrandom Sampling

Nonrandom sampling occurs when the process of sample selection is not completely objective, and where some form of judgment or subjectivity is involved. Even though it is advisable to take random samples wherever possible, nonrandom samples are sometimes unavoidable. **It is important to realise, however, that assertions about the population based on nonrandom samples may be invalid**. For example, we saw in Figure 6.7 that judgment sampling led to biased estimates of the mean weight of the wood population. Generally, experience has shown that judgment samples often tend to be biased upwards.

> **Q27.** *Explain why the sample selected in the "Should New Zealand become a republic?" survey is a nonrandom one.*

We now briefly consider three particular methods of nonrandom sampling, although in practice there are many other ways in which nonrandom samples may be selected in particular situations.

☞ *Quota Sampling*

A quota sample is one that is representative by various factors, such as age, sex, and socioeconomic status. The person taking the sample has freedom to choose a specified number (or quota) of people fitting a certain profile, such as a given sex, age, and social class. Quota sampling is essentially the nonrandom counterpart to stratified sampling, where various groups are judgmentally, rather than randomly, sampled.

Quota sampling is often used in market research surveys. The advantage of this method is that it is simple and inexpensive to carry out. However, since the method of selection is not random, it can give an unrepresentative sample since the interviewer's personal biases, conscious or unconscious, are used in selecting sample members.

A market researcher is sampling customers leaving an up-market retail store. Her quota consists of interviewing a certain proportion of males and females of different ages and employment status. The profile of the selected subject is determined from the answers to the first few questions.

> **Q28.** *Why might her results be biased?*

✍ Judgment Sampling

Here we do our conscious best to get a "fair" sample by picking those items that, in our judgment, form a representative cross section of the population. However, it is almost impossible for any person to be totally objective and there is a serious risk of distortion due to the sampler's personal prejudices, as the wood experiment clearly demonstrated.

✍ Self-Selected Sampling

Here, inclusion in the sample is left up to the members of the population, who each decide whether or not they will be involved. Only those respondents who are sufficiently motivated to reply are included. This is often seen on "talkback" TV or radio programmes, where the viewer or listener is asked to phone in about an area of topical interest. This method encourages people with strong opinions to participate and for this reason it is unlikely that the sample will be representative of those with more moderate opinions.

> **Q29.** *Over 4000 people rang a number on a TV talk show to express their vote for or against changing the name Hamilton to Waikato City. Over 80% were opposed to the change. What conclusions do you draw from this survey?*

6.8   The Accuracy of Sample Estimates

There are many reasons why the data produced by a sample could be inaccurate, even worthless. Even if a random sample has been used, the sample may not be representative of the entire population simply on account of the chance element involved in random sampling. This sort of inaccuracy is known as *sampling error*, which will be explored in more detail in the next chapter. Sampling errors are unavoidable, but the likely size of these errors can be estimated.

However, there may be other sources of inaccuracy that are not the result of sampling error. These are *nonsampling errors*, which could introduce systematic errors into the estimation of population parameters. Nonsampling error leads to bias in the estimation process, and taking a larger sample of data will not necessarily reduce this bias. Every effort should be made to avoid nonsampling error, mainly because you can never know how large the bias might be.

There are many reasons why nonsampling errors can arise, for example:

- Nonrandom sampling methods are used. Judgment introduces subjectivity, which can cause bias.
- There is a high nonresponse rate. This is a form of self-selection (or non-selection), and it is possible that those who choose to respond will have a different opinion from those who do not. It is often better to have a smaller sample size and use the money saved to increase the response rate.
- The target population is not clear, there is no adequate sampling frame, or the list is badly out of date.
- In a survey there are ambiguous, misleading, or biased questions, or the interviewer is biased.

6.9   Data for the Wood Experiment

The weights of the population of 100 blocks of wood, or equivalently the weights of the blocks in Figure 6.3, are given in Figure 6.8.

6.10   Exercises

1. A lawn mower manufacturer is concerned that the new design of the starting mechanism has made starting the mower more difficult. The company selects a random sample of 40 people who weigh less than 50 kg, and each person attempts to start the mower. The number of attempts needed to start the mower is recorded. This is an example of
   a. simple random sampling
   b. stratified sampling
   c. quota sampling
   d. cluster sampling
   e. none of the above

2. A company conducts an advertising campaign on several television channels and radio stations. Consumers ringing the free phone number are asked to

Block Number	Weight (gm)	Block Number	Weight (gm)	Block Number	Weight (gm)	Block Number	Weight (gm)	Block Number	Weight (gm)
1	45	21	40	41	45	61	85	81	30
2	105	22	30	42	35	62	25	82	30
3	30	23	10	43	65	63	85	83	15
4	30	24	35	44	40	64	10	84	30
5	50	25	45	45	15	65	55	85	10
6	15	26	50	46	40	66	10	86	45
7	25	27	30	47	40	67	30	87	10
8	90	28	30	48	25	68	60	88	45
9	75	29	35	49	40	69	25	89	10
10	50	30	35	50	45	70	70	90	60
11	40	31	10	51	45	71	50	91	15
12	45	32	30	52	15	72	25	92	80
13	50	33	70	53	35	73	15	93	50
14	75	34	40	54	45	74	10	94	5
15	30	35	50	55	25	75	35	95	20
16	50	36	35	56	40	76	40	96	55
17	50	37	60	57	45	77	30	97	10
18	55	38	10	58	25	78	40	98	30
19	45	39	10	59	70	79	15	99	55
20	40	40	55	60	65	80	70	100	45

Figure 6.8   Weights of the Population of Wood Blocks

identify the medium through which they found out about the product. The manager received 4500 replies to the campaign. He plans to use the replies to work out where to advertise for future campaigns. Which one of the following is true?

a. it is a simple random sample
b. it is a stratified sample
c. it is a cluster sample
d. it is a quota sample
e. it is a self-selected sample

3. A major advantage of a random sample compared with a nonrandom sample is that
a. it prevents destructive sampling
b. it is less expensive
c. sampling error can be estimated
d. it saves time

4. Any characteristic of a population is called a
a. variable
b. attribute
c. parameter
d. standard error
e. statistic

5. A news television programme asks viewers to phone or fax in their agreement/disagreement with the expulsion of school pupils for using an illegal substance at school. Suppose that the television station received 10,000 replies. Which of the following is true?
   a. This is a large random sample.
   b. The large sample will provide results that are representative of the population.
   c. The results are likely to be biased.
   d. All of the above.
   e. None of the above.

6. Suppose that British Airways is interested in carrying out a market research survey to find out its customers' views on its catering services.
   a. How should this customer survey be carried out?
   b. How should they sample their customers?
   c. What type of questions should be asked (e.g., open ended, yes/no types, using 5-point scales) and why?

7. On September 17, 1997 a referendum was conducted in Wales on the question of parliamentary devolution. People on the electoral register in Wales were eligible to vote in this referendum. They were asked to agree or disagree with the following proposition: "I agree there should be a Welsh assembly." The result, from a turnout of 50.1%, was as follows:

Yes	559,419	50.3%
No	552,698	49.7%
Majority in favour	6,721	0.6%

   Comment on the outcome of this referendum.

8. In 1999, 237 students from a Management Statistics course handed in feedback forms at the end of the course. To analyse the open-ended questions, a systematic sample was taken. A sample of about 50 was the maximum that could be processed.
   a. Define the sampling frame and describe how to take a systematic sample from the population of forms.
   b. The forms are stored by tutorial group, with about eight students per group. Do you think that this ordering of the forms would make the sampling method appropriate or inappropriate? Justify your answer.

9. The manager of North Island Airlines has recently purchased health insurance for the company's employees. They want to find out the average weekly medical expenses for the employees. There are 397 maintenance staff, 614 cabin staff, 162 senior pilots, and management staff. A sample size of 100 is to be used.
   a. Which sampling method would you use to estimate the mean weekly medical cost per employee?
   b. Explain how you would take the sample.

10. State whether each number in bold in the exercises below is a parameter or a statistic.

a. The National Statistics Office last month interviewed 25,000 members of the labour force of whom **8.6%** were unemployed.
b. A consignment of widgets has a mean weight of **15.4 g**. This is within the specifications for acceptance of the consignment by the purchaser. However, the purchaser falsely rejected the consignment because a randomly chosen sample of 50 widgets had a mean weight of **16.2 g**, which is outside the specified limits.
c. A telephone sales company based in Swindon uses a device that dials residential phone numbers in that city at random. Of the first 200 numbers dialed, **19** are unlisted. This is not surprising, because **8%** of all Swindon phone numbers are unlisted.
d. A market researcher investigating the reactions to a new slimming food conducts a comparative experiment with 20 randomly chosen subjects. A control group of ten is fed a placebo, while the experimental group of ten is fed the new product. After 4 weeks, the mean weight loss is **14 kg** for the control group and **25 kg** for the experimental group.

11. Jenny, the marketing manager of the telemarketing company EazySell, wants to determine the average number of telephone calls her operators make each month. There is little variation in the number of calls within a month, but considerable variation from month to month. Jenny also suspects that they make more calls earlier in the week than later. Explain carefully what sampling method Jenny should use in order to be able to accurately estimate the overall average number of calls made monthly.

12. In the week preceding a local election, the editor of the *Loughborough Echo* decides to take an opinion poll to try to predict the outcome of the election. Four possible methods are proposed for selecting a sample of voters.
    1. Interview every 20th person over 18 entering a local supermarket between 09:00 and 17:30 this coming Friday.
    2. Select 3 streets at random from the Loughborough street directory and interview every person over 18 in each household. These interviews will be conducted over the next four evenings.
    3. Select every 10th page from the local residential telephone directory and call the first number with a Loughborough prefix (01509) listed on each page. The calls are to be made over the next 4 evenings, and information collected from whoever answers the phone, provided they are over 18.
    4. Two interviewers will interview people in the Curzon cinema lobby this coming Saturday between 12:00 and 22:00. One interviewer will select only males, and the other only females. They will interview as many people over 18 as they can in the time available.
    a. Classify each of the proposed sampling schemes as either random or nonrandom.
    b. Classify each of the random schemes as simple random, stratified, systematic, or cluster.
    c. Classify each of the nonrandom schemes as quota, judgment or self-selected.
    d. Give a possible source of bias associated with each sampling scheme (that is, a reason why it may not produce data that are representative of the voting population).

Chapter 7

Estimation

7.1 Sampling Error

We have seen in earlier chapters that data are usually collected in order to gain *information* about, and better understand, some particular population or process. In some circumstances it may be appropriate to attempt to collect data about every member of the population (a *census*), but this is often impractical because many of the populations we wish to study are impossibly large. It is therefore more usual to collect information from only a sample of the population, as described in the previous chapter. In practice, there are many ways in which samples of data may be collected, both random and nonrandom, but however the sample is taken, it provides data for only a portion of the entire population. So, it is unlikely that a sample will give precise information about the population. There will usually be error from using a sample to estimate population characteristics. With a random sample, this error is known as *sampling error*. With a nonrandom sample, however, estimates may also be *biased*. As well as sampling error, there may be a systematic error caused by the element of judgment used in the sampling process.

Whenever we rely on sample information it is important to have an appreciation of how large the sampling error might be. Without an understanding of the sampling error, we are likely to misuse the information and possibly make wrong decisions.

 Value Added Tax (VAT)

The Tax Department sampled 350 VAT returns and found that 43 had significant errors in them.

Q1. *What can you say about the proportion of all VAT returns with significant errors?*

Q2. *What other information would you like to know?*

 ## Union Membership

A random sample of 50 small businesses in a city of 200,000 inhabitants found that in only 12 of them were any of their workers members of a trade union.

Q3. *What conclusions can you draw from this sample about trade union membership in small businesses in that city?*

Q4. *What conclusions can you draw about trade union membership nationally?*

 ## Advertising Awareness

An advertising agency has been commissioned to develop a TV advertising campaign for a major bank. The campaign involves two new advertisements appearing at peak viewing times between 19:30 and 20:00 each evening. After the first screening of each advertisement, the agency surveys its viewers' panel of 386 randomly chosen people to determine whether they were watching TV at the time, and if so their awareness of the advertisement by being able to recall the name of the bank concerned. They obtained the results shown in Figure 7.1.

Advertisement	Number watching TV	Number aware of advertisement
A	228	155
B	216	121

Figure 7.1 Advertising Awareness Data

> **Q5.** *How would you describe the viewers' awareness of each advertisement?*
> **Q6.** *Is there any real evidence that one of the advertisements has had more impact than the other?*

 ## Thorcam Domestic Appliances

One of the key components in Thorcam washing machines is the timer that controls the washing cycle. In recent months there have been more customer claims than usual for timers failing within the 1-year warranty period. Such claims result in an expensive replacement or repair.

To assess the life of timers under normal operating conditions, a sample of 25 is put on test and run until failure. The number of hours of continuous operation until failure is recorded in Figure 7.2. The average (mean) life is 686 hours and the standard deviation is 224.4 hours.

Timer	Life	Timer	Life	Timer	Life	Timer	Life	Timer	Life
1	470	6	842	11	684	16	694	21	658
2	702	7	1020	12	696	17	516	22	898
3	814	8	968	13	580	18	400	23	670
4	98	9	950	14	360	19	912	24	654
5	548	10	988	15	594	20	900	25	534

Figure 7.2   Life of Thorcam Washing Machine Timers (hours)

> **Q7.** *What can you conclude about the mean lifetime of the timers?*

Wood Experiment

In the wood experiment in Chapter 6, students are asked to select 2 samples of 10 blocks of wood, the first using their judgment and the second using a random sampling approach. The mean block weights from the random samples of a group of 171 students are shown as a bar chart in Figure 7.3.

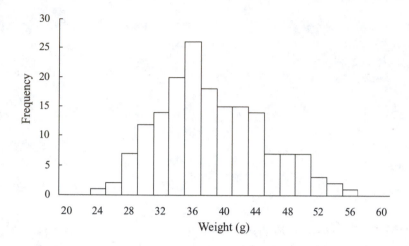

Figure 7.3   Wood Experiment—Random Sample Means

In practice, if we had to estimate the mean weight of the wood population, we would take just 1 sample of blocks, rather than repeat the experiment 171 times. However, the distribution of results in Figure 7.3 tells us what range of mean values we are *likely* to get from a random sample of 10 blocks of wood.

Q8. From the data in Figure 7.3, what are the minimum and maximum values that we could realistically get for the mean weight of a random sample of 10 blocks of wood?

Q9. Would it be the same if we took a sample of 40 blocks of wood?

These five examples are each concerned with the accuracy of sample information when estimating some population *parameter*. In other words, how large is the sampling error? However, what we are estimating is different in each case. In the first three examples, we have attribute data and wish to estimate the *proportion* of cases in which that attribute occurs, namely, the proportion of VAT returns with significant errors, the proportion of small businesses with trade union members, or the proportion of viewers who are aware of an advertisement. In the fourth and fifth examples, we have interval data and our concern is with the population *mean*, namely, the mean life of timers or the mean weight of blocks of wood. We are usually interested in estimating a proportion or a mean. However, we might be concerned with other population parameters, for example:

- the population median; for example the median traveling time to university,
- the population standard deviation; for example of the volume of coke in a can, or

- the population mean for a value of another variable; for example average salary of accountants at age 40.

In this chapter we will concentrate on sampling errors in estimates of a population proportion and a population mean.

7.2 Estimating a Proportion

To understand the concept of sampling error, it is important to realise that a given sample result is only *indicative* of the true situation, and will not necessarily reflect exactly the situation in the population as a whole. In the awareness data, we cannot conclude that exactly 68% of the viewing population were aware of advertisement A just because our sample gave that result. If we took another sample of 228 viewers, we could get (say) 72% of this new sample being aware of the advertisement. A further sample of 228 might yield a proportion of 65%, and so on. The extent to which sample results, such as these, can vary from the true figure is called the *margin of error*.

 ### *An Opinion Poll*

The following statement is from a local newspaper just prior to a national election.
  "In an opinion poll of 1000 electors, one of the political parties has a support of only 2%. This is below the margin of error (of 3%)."

> **Q10.** *What do you think this statement implies?*

In the advertising awareness example, as you will see later, the proportion of the sample of viewers that are aware of advertisement A should be within about 6% of the true figure for the population as a whole. This means that we have a margin of error of ±6% in our sample estimate. That is, our sample proportion could be in error (different from the population proportion) by as much as 6%.

To calculate the margin of error we could, in principle, take a succession of random samples of 228 viewers and see how much variation there is in the sample proportions we get. But this is not necessary because statistical theory can be used to infer what this variation would be. In Section 5.5 we saw that a sample proportion is approximately normally distributed, provided the sample is reasonably large, with a standard deviation given by

$$\sqrt{\frac{p(1-p)}{n}}$$

where p is the expected proportion, that is, the true proportion for the entire population.

With this result, we can use the theory of the normal distribution to determine the margin of error. For example, 95% of the values in a normal distribution fall within 1.96 standard deviations either side of the mean value, and so 95% of sample proportions will fall within a *margin of error* given by

$$\pm 1.96 \sqrt{\frac{p(1-p)}{n}}$$

In other words, 95% of the time, the difference between a sample result and the true value will be less than this margin of error.

Advertising Awareness

In a random sample of 228 people who were watching TV when advertisement A was shown, 68% were aware of the advertisement.

> **Q11.** *Verify that the 95% margin of error for the awareness of advertisement A is about 6%.*
>
> **Q12.** *Do you think that the margin of error for the awareness of advertisement B will be the same? Calculate the margin of error for advertisement B.*
>
> **Q13.** *What can you conclude from these results about a comparison between advertisements A and B in terms of their relative awareness?*

An Opinion Poll

Think again about the quote from the newspaper regarding the political party having 2% support in an opinion poll of 1000 electors.

> **Q14.** *Does the statement make sense?*

Q15. *What should the margin of error be in this case?*

Q16. *Where do you think the figure of 3% has come from?*

 ## Confidence Interval

When specifying a margin of error we are really saying that there is a range, or interval, around the sample result, within which the population parameter *probably* lies. This interval of values either contains the population parameter, or it does not, but we have a certain level of belief, or *confidence*, that the population parameter falls within this interval. For this reason, the interval is called a *confidence interval*.

Of course, there is nothing absolute about a confidence level of 95% that we have used. In some cases we may want to be more sure of our conclusions, and so choose to work with (say) a 99% confidence level, or possibly even higher than that. Alternatively, we might be content to accept a lower level of confidence in our conclusions, although it is unusual for the confidence level to be less than 90%.

Q17. *How does the margin of error alter if the confidence level increases or decreases?*

Q18. *What value do we use in place of 1.96 for 90% confidence? For 99% confidence?*

Q19. *Could we be absolutely certain by using a confidence level of 100%? Explain why or why not?*

7.3   Sample Size for Proportions

The formula for the margin of error of a proportion shows that it is inversely proportional to the sample size n. That is, as the sample size increases, the margin of error decreases.

Q20. *Is this what you would expect? Why, or why not?*

As we have seen, for any confidence level the margin of error for a proportion p can be determined from the sample size. Conversely, by rearranging the formula for the margin of error, we can determine the sample size required for a specified margin of error, namely

$$n \geq \left(\frac{z \times \sqrt{p(1-p)}}{\text{margin of error}} \right)^2$$

where z is taken from Figure 4.21 (reproduced below as Figure 7.4) for the desired confidence level.

Usually p will be an unknown quantity, as we are determining the sample size required to estimate p with a desired margin of error. When p is completely unknown, you should assume the worst-case scenario and use $p = 0.5$ because, as we saw in Figure 5.17, this is where the standard deviation, and hence the margin of error, is greatest. If p can be roughly estimated within a particular range, you should use the value of p nearest to 0.5. For example, if we can be fairly sure that p is somewhere between 0.1 and 0.3, the value $p = 0.3$ should be used as this will lead to the largest value of n.

Note that for a 95% confidence level $z \approx 2$, so that $z\sqrt{p(1-p)} = 1$ when $p = 0.5$.

This gives the simple result: $n \geq 1/(\text{margin of error})^2$ for 95% confidence in the worst-case situation.

Confidence level	z
90%	1.645
95%	1.960
98%	2.326
99%	2.576

Figure 7.4   Selected Percentage Points of the Standard Normal Distribution

Q21. *In a political opinion poll a 95% margin of error of ±3% is required. Show that a sample of about 1000 electors is needed to estimate the percentage vote for any party.*

Q22. *What would be the 95% margin of error, using a sample size of 1000 electors, for a political party with a level of support about 5%?*

7.4   Estimating a Mean

We shall now consider the problem of estimating the mean of a population from a random sample of interval data. As with proportions, there is sampling error involved in any estimate of a population mean, but how do we quantify it? For example, how accurate is the mean life of the sample of 25 timers in Figure 7.2 as an estimate of the mean life of all timers? To answer this question we need to know how much variation there will be in the distribution of a (large) number of such sample mean values.

This is exactly what we did in the wood experiment. We took a number of random samples of size 10 from the population of blocks of wood, and calculated the mean of each sample. We then constructed a bar chart of these means, an example of which is shown in Figure 7.3. If we have a large enough number of samples, we could approximate this plot by a smooth distribution curve. This distribution is called the *sampling distribution of the mean*.

✍ *Sampling Distributions of Mean Weights*

How does the sampling distribution of the mean compare with the distribution of the population from which the samples were taken? In the wood experiment, we know that the population consists of 100 blocks with weights ranging from 5 to 105, as shown by the bar chart in Figure 7.5.

Q23. *How does the average, spread, and shape of the sampling distribution of the mean (Figure 7.3) compare with the population distribution (Figure 7.5)?*

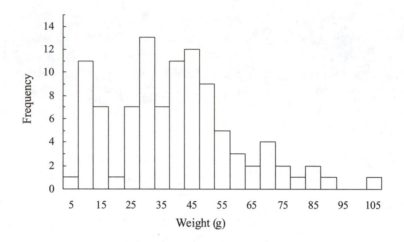

Figure 7.5   Distribution of Weights of 100 Blocks

Q24. What do you think would happen to the sampling distribution of the mean in Figure 7.3 if we were to take smaller samples (say 5) or larger samples (say 20)?

Using random sampling methods, we can generate sampling distributions corresponding to different sample sizes. To give a direct comparison with Figure 7.3, we have taken 171 independent random samples of sizes 5 and 20 from the population of 100 blocks shown in Figure 7.5. The mean weight of the items in each sample was calculated, and the 171 sample means are shown in the bar charts in Figure 7.6, together with the chart for $n = 10$.

Q25. What can you say about the location, spread, and shape of these distributions?

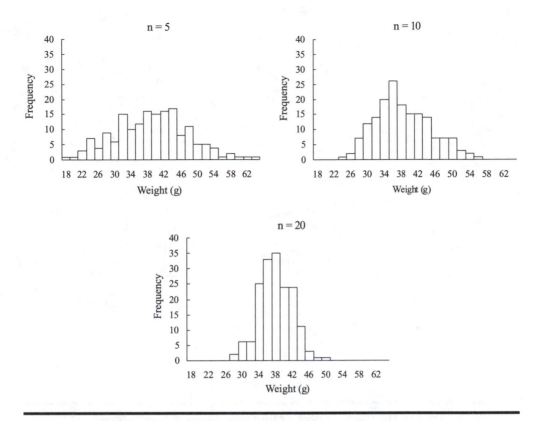

Figure 7.6 Sampling Distributions of the Arithmetic Mean for *n* = 5, 10, and 20

✍ *Properties of the Sample Mean*

Figure 7.6 shows how the sample mean ($\bar{x}$) varies for different sample sizes *n*. These distributions illustrate the following properties of the sampling distribution of the mean.

- The sampling distribution of the mean tends to be *bell shaped* and approximately normally distributed, irrespective of the shape of the distribution from which the sample of data is taken. The normality of the distribution improves as the original distribution itself becomes more normal, and as the sample size increases. In practice, it is safe to assume that the distribution of $\bar{x}$ will be normal if
  - ➢ the sample size *n* > 30, or
  - ➢ *n* > 10 and the distribution of the observations is reasonably symmetric, or
  - ➢ the distribution of the observations is close to normal, for any value of *n*.
- The mean of the sampling distribution is the same as the population mean μ, for any value of *n*.
- The *standard deviation* of the sampling distribution is $\sigma/\sqrt{n}$, where σ is the standard deviation of the population. The standard deviation of the sampling distribution is usually called the *standard error* of the mean.

For the population of weights in Figure 7.5, the mean of all 100 items is actually 39.4 and the standard deviation is 20.63. So, if samples of size 5 are taken

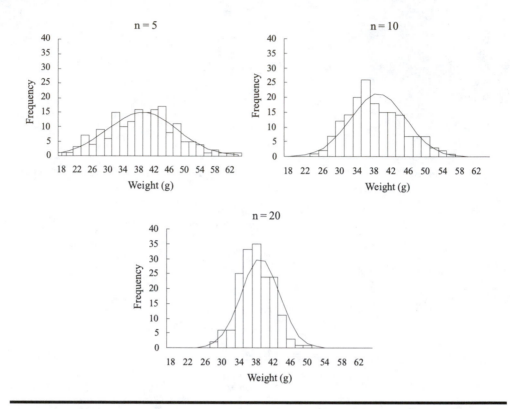

Figure 7.7   Sampling Distributions of the Mean with Normal Distributions

from this population, the sample means should average 39.4, with a standard deviation of $20.63/\sqrt{5} = 9.23$. Alternatively, with samples of size 10, the standard deviation should be $20.63/\sqrt{10} = 6.52$, and for samples of size 20 it should be $20.63/\sqrt{20} = 4.61$.

Normal distributions with a mean of 39.4 and with the appropriate standard deviation are shown in Figure 7.7 superimposed on the relevant bar chart from Figure 7.6. It can be seen that these normal models represent the data well, despite the fact that the sample sizes (5, 10, and 20) are not particularly large.

✎ *Margin of Error for a Mean*

As previously mentioned, we do not expect the sample mean to be exactly equal to the population mean μ. Therefore, we need to provide information about the accuracy of our estimate. As with proportions, this is accomplished by stating a *margin of error* in our estimate for the unknown population parameter (the population mean).

The margin of error gives us

- an interval of uncertainty within which the population mean *probably* lies, and
- some idea of how often the population mean would be expected to be within this interval in repeated sampling. We specify the strength of "probably" in terms of a confidence level.

Knowing that a sample mean is approximately normally distributed, with a standard deviation of $\sigma/\sqrt{n}$, the margin of error is

$$\pm z \frac{\sigma}{\sqrt{n}}$$

where z is an appropriate percentage point of the standard normal distribution from Figure 7.4.

This is essentially the same formula as in the case of proportions, but with a different expression for the standard deviation. That is, instead of $\sqrt{p(1-p)/n}$ we use the standard error $\sigma/\sqrt{n}$.

This margin of error is valid when:

- a random sample of observations has been taken,
- the population standard deviation (σ) is known, and
- the sampling distribution of the mean is approximately *normal*, that is, the population being sampled is itself approximately normal, or the sample size is sufficiently large.

Recall that the standard deviation of the weights of items in Figure 7.5 is 20.63. The 95% margin of error for the mean of a sample of 10 items is therefore

$$\pm 1.96 \times \frac{20.63}{\sqrt{10}} = \pm 12.79$$

Q26. *Explain in simple terms what this margin of error means.*

Q27. *Would you expect that the 90% margin of error for a sample of 5 items would be more or less than 12.79? Verify your answer by performing the calculation.*

Q28. *Remembering that the mean of the population is actually 39.4, how likely is it that the mean of a random sample of 10 items would be less than 25?*

Thorcam Domestic Appliances

Recall the problem at the beginning of this chapter concerning the life of Thorcam washing machine timers. A random sample of 25 timers, when tested to failure, gave an average life of 686 hours, with a standard deviation of 224.4 hours.

> **Q29.** *Assuming that the standard deviation of the population of all timers is about 220 hours, calculate the 95% margin of error for this sample mean of 686 hours.*
>
> **Q30.** *Explain in plain language what this means.*

7.5   Sample Size for Means

We can again use the formula for the margin of error in a sample mean to determine how large a sample is necessary to produce a required margin of error. Rearranging the formula for the margin of error we get

$$n \geq \left(\frac{z \times \sigma}{\text{margin of error}} \right)^2$$

For example, how many timers need to be tested to estimate the mean lifetime of all timers to within 40 hours? If we assume a 95% margin of error, then $z = 1.96$, so that

$$n \geq \left(\frac{1.96 \times 220}{40} \right)^2 = 116.2, \quad \text{i.e., 117 timers}$$

> **Q31.** *Suppose we want to be 95% confident that the sample estimate will be within 20 hours of the population mean. What sample size would be required?*

> **Q32.** *What general conclusions can you draw about the relationship between the margin of error and the sample size required?*
>
> **Q33.** *In particular, to halve the margin of error, we must increase the sample size by what factor?*

In practice it is very hard to get an accurate estimate of the standard deviation σ, so the sample size we calculate is usually, at best, only approximate. But it is better to have a general idea of how big a sample you might need to take, and the costs associated with it, before the event. It might be appropriate in some cases to take a small sample to get a better estimate of the population standard deviation and use this to determine the sample size more accurately. The initial sample can then be "topped-up" to the required number. Of course, other factors influence sample size, such as cost and available resources. However, experience suggests that people tend to overestimate the sample size they require. Unnecessary money and resources can be wasted collecting data from very large samples.

7.6 Unknown Population Standard Deviation

In practice, we seldom know the population standard deviation σ. In this case there are two changes to the margin of error.

1. As you might expect, we now use the sample standard deviation (s) in the formula to replace σ.
2. As a result, because s varies from sample to sample, the margin of error will be larger. To allow for this, in the formula for the margin of error we replace the z-value we looked up from the standard normal distribution (Figure 7.4) with a value from Student's t-distribution.

Student's t-distribution, or simply the *t-distribution*, is another commonly used statistical distribution. It is also a symmetric distribution, and has one parameter called the *degrees of freedom*. The concept of degrees of freedom was discussed in Section 4.4 in relation to the standard deviation, where the number of degrees of freedom was $n - 1$. The number of degrees of freedom in the calculation of s is simply an indicator of how good s will be as an estimator of σ. The bigger the sample, the more degrees of freedom it will have, and the better s will be as an estimate of σ. It is this quantity that determines the value of t we use to replace z in our formula for the margin of error. For a particular confidence level, t will always be greater than z, but as the degrees of freedom get larger, t gets closer and closer to z.

The margin of error for the population mean μ, based on a random sample of size n, is

$$\pm t_{n-1}\frac{s}{\sqrt{n}}$$

where t_{n-1} is an appropriate percentage point of the t-distribution with $n-1$ degrees of freedom.

This margin of error is valid provided that:

- the n observations are a random sample from a *normal* population, and
- the unknown population standard deviation is estimated by the sample standard deviation s.

A table of selected percentage points of the t-distribution is in Appendix B.2. The table gives the value of t such that a specified percentage (P) of the distribution lies between $-t$ and $+t$, for P equal to 90, 95, 98, or 99 and for different numbers of degrees of freedom (df). In other words, the appropriate row of Appendix B.2 gives us the required percentage points instead of the values in Figure 7.4.

 ## Thorcam Domestic Appliances

In our previous calculation to determine the margin of error for the life of Thorcam washing machine timers, we made the (perhaps rather bold) assumption that the standard deviation of the population of timers was about 220 hours. However, in reality, we do not know what the population standard deviation is, and it may **not** be 220. However, using the t-distribution, we can now work out an exact margin of error using the *sample* standard deviation $s = 224.4$, with $25 - 1 = 24$ degrees of freedom.

For a 95% confidence level, $t_{24} = 2.064$, giving a margin of error

$$\pm 2.064 \times \frac{224.4}{\sqrt{25}} = \pm 92.6 \text{ hours}$$

With 95% probability, the sample mean (686 hours) should be accurate to within 92.6 hours. The mean life of all timers should, therefore, be within the range $686 - 92.6 = 593.4$ hours to $686 + 92.6 = 778.6$ hours.

7.7 Finite Populations

So far we have implicitly assumed that the population being sampled is sufficiently large that any sample we take constitutes a tiny, insignificant fraction of the whole population. What if this were not the case? Is there a difference between a sample of 50 from a population of millions, and a sample of 50 from a population of 100?

> **Q34.** *What do you think? Does the population size affect the margin of error? If so, how?*

To see that population size does matter, you only have to think of the extreme case, when we sample the entire population. For example, if we take a sample of 100 items from the wood population in Figure 7.5, we have observed every item in the population and so any sample statistics we calculate (such as the mean) will be totally accurate, with no error. Whenever we take a sample from a finite population, where the sample constitutes a non-negligible fraction of the population, we must amend our margin of error formula by multiplying it by the *finite population correction factor* given by

$$\sqrt{1 - \frac{n}{N}}$$

where N is the size of the population. The correction factor is the square root of the fraction of the population that is *not* sampled.

Wood Experiment

Suppose we take a random sample of 10 items from the population in Figure 7.5, obtaining the following values:

$$10, 35, 40, 20, 75, 60, 40, 55, 10, 85$$

> **Q35.** *Assuming we know that the standard deviation of the population is 20.63, calculate a revised 95% margin of error incorporating the finite population correction factor.*
>
> **Q36.** *Assuming we do not know what the population standard deviation is, estimate this by calculating the sample standard deviation and then determine the new margin of error.*

7.8   What Affects the Margin of Error?

The various formulae that we have developed in this chapter show that the margin of error is dependent generally on three key factors of the sampling situation in question, namely

1. The amount of variation in the population being sampled. The more variable the population, the greater will be the margin of error in any sample estimate.
2. The size of the sample. The larger the sample, the smaller the margin of error. The main point to note here is that sampling error is related to the *square root* of the sample size, and so to halve the error, we must quadruple the sample size.
3. The chosen level of confidence reflected in the particular percentage point (z or t) used. A greater confidence level (more confidence in the conclusion) will increase the resulting margin of error.

As described in the previous section, where the population being sampled is small, then the fraction of the population that is sampled also has an effect, but this is seldom important as populations are usually large, if not infinite.

The first factor above is not usually controllable, as it is inherent in the population or process being studied. So, to reduce the margin of error, it is necessary to either increase the sample size, or to sacrifice some confidence in the results. If the sample size is also fixed or predetermined, then decreasing the level of confidence is the only way of reducing the margin of error.

7.9   Tests of Hypotheses

We collect data to find out about a population or process. In the Thorcam Appliances example, we wish to know the long-run average life of timers. In other words, how long can we *expect* any timer to last until breakdown? We can conduct an experiment or test, and use the results from this experiment to estimate the population mean with an appropriate margin of error.

Sometimes, however, we have an idea, based perhaps on some theory, experience with similar processes, or a hunch, of what the population mean should be. Then we might be interested in seeing whether the results of our experiment are consistent with this idea. For example, as a result of the increased number of warranty claims, we may have reason to question whether the performance of Thorcam timers has changed from the previously established level. This could be the result of a design change, or changes to the manufacturing process, or a different type of washing machine in which the timers are now installed.

 Thorcam Domestic Appliances

The supplier of the timers used in Thorcam's washing machines claims that their timers will function for at least 750 hours.

Q37. *Do you think that the results given in Figure 7.2 cast doubt on the claim of the supplier? Why?*

What we need to do is to carry out a *test of significance*, or *test of hypothesis*. Our hypothesis in the Thorcam example is that the population mean is 750. If we calculate a confidence interval, and find that our hypothesised mean (750) does not fall within this interval, then the evidence suggests that our hypothesis is unreasonable, or equivalently that there is a *statistically significant* difference between the sample mean and the hypothesised population mean. As we saw earlier, the value of 750 falls comfortably within the calculated confidence interval, and so there is no firm evidence to reject a mean value of 750. Of course the problem might be that we have not taken a large enough sample to detect a difference, and we might decide to collect some more data to see if this is the case.

✌ *Margin of Error and Tests of Hypotheses*

The concepts involved in testing hypotheses are similar to those behind the margin of error. We shall not develop them further here. In business the problem of estimation is far more important, or should be, than testing some hypothesis. The primary purpose of collecting data should be to find out about the population or process under study, not to test whether some theory or other is valid.

However, hypothesis tests can sometimes be a useful decision tool, but you should bear in mind the following general points about margins of error and hypothesis tests.

A margin of error:

- gives *quantitative* information about the range of uncertainty in some population or process parameter, and
- demonstrates the *accuracy* of some estimate by the size of the margin of error.

Hypothesis testing:

- gives *qualitative* information about the correctness of some belief regarding a population or process parameter, but it
- cannot totally confirm or deny such a belief, and it
- can be fallible when there is a great deal of variation in the process, as you may be misled into thinking that the data are consistent with the hypothesis when really there is too much variation to be able to tell.

Brian Joiner, an eminent statistician and management consultant, says "I would say never test a hypothesis unless you're desperate and can't think of anything else to do. When you're testing a hypothesis, just knowing whether that hypothesis is to accept or reject is not a very informative statement. It doesn't give you useful

information as a basis for action... It's a big mistake for statisticians to be teaching those kinds of things. It's doing serious damage." (*Quality Progress*, 1988, pp. 31–32.)
Hence our advice is to use *estimation with a margin of error.*

7.10 Exercises

1. For random samples of size 100 from a population with mean 92 and standard deviation 25, the range of the middle 50% of sample means is approximately
   a. 67 to 117
   b. 87 to 97
   c. 89.5 to 94.5
   d. 90.3 to 93.7
   e. 91.7 to 92.3

2. The production manager wants to evaluate the lifetime of the latest batch of light bulbs. He obtains a random sample of 15 light bulbs and tests them until failure. It is known that the lifetimes are normally distributed. In the construction of the margin of error for the sample mean, it is appropriate
   a. to use the normal distribution only if the population standard deviation is known
   b. to use the sample standard deviation and the normal distribution
   c. to use the sample standard deviation and the *t*-distribution
   d. either (a) or (c)
   e. either (b) or (c)

3. A political opinion poll was conducted using a random sample of 1000 people. A 99% margin of error was calculated for the proportion of voters in favour of the present government. Suppose that we want to halve the margin of error. Then we must
   a. double the sample size
   b. decrease the sample size by half
   c. multiply the sample size by four
   d. divide the sample size by four
   e. none of the above

4. Sothany Ngyuen, marketing manager of EasyGo credit card company, is interested in the proportion of customers who pay their accounts in full each month. If he wants to estimate the proportion to within 4% with 95% confidence then the sample size required is
   a. 93
   b. 151
   c. 601
   d. 2401
   e. none of the above

5. Assuming that the population standard deviation is known, what happens to the size of the margin of error for estimating the population mean when the level of confidence increases from 90% to 99%?
   a. it increases by 9%
   b. it decreases by 9%

    c. it increases by about 31%

    d. it decreases by about 31%

    e. it increases by about 57%

    f. it decreases by about 57%

6. A bank manager is interested in the proportion of customers who have multiple accounts at the bank. A random sample of 500 customers is taken showing that 195 of these customers have multiple accounts.

    a. Calculate a 95% margin of error for the proportion of customers having multiple accounts.

    b. Describe in plain language what this means.

7. It is common practice, prior to most local and national elections, for the media to predict the winning party. Let us suppose that there are just two parties, Democrat and Republican, contesting an election. In a random sample of 353 voters, 205 (58%) said that they intended to vote for the Democrat Party.

    a. What can you deduce from this information about the likely outcome of the election?

    b. How confident would you be that the Democrats will win?

8. The Healthy Breakfast Company carried out an advertising campaign on a new rice-based breakfast cereal, Extra Fruity. The manager would like to determine the proportion of students who had Extra Fruity for breakfast at least once during the week following the campaign. From a systematic sample of 200 students, he found that 18 of them had eaten this cereal for breakfast at least once in the week concerned.

    a. Construct a 99% margin of error for the proportion of students eating the cereal at least once in the week in question.

    b. The agency planning the campaign was targeted to achieve at least 15% penetration in the young consumer market. What do the sample results tell you regarding this objective?

9. The manager of a savings bank wants to estimate the average amount held in passbook savings accounts by depositors at her branch. A random sample of 25 depositors is taken. The sample mean is $4750 and the sample standard deviation $1200.

    Assuming that the amount in passbook savings accounts is normally distributed, calculate a 95% margin of error for the mean amount in passbook savings accounts.

10. The volume dispensed into soft drink bottles is known to be approximately normally distributed with a standard deviation of 0.05 litres. A random sample of 50 bottles gives a sample mean volume of 1.99 litres. Bottles that contain less than the specified volume (2 litres) can lead to the manufacturer being prosecuted.

    a. What is the 90% margin of error for the mean volume? Comment on your results.

    b. The manager wants to estimate the mean volume to within 0.01 litres with 95% confidence. How many observations should be included in the sample?

11. A university wants to determine the average income their students earn during the summer.

a. Explain briefly how you might choose a stratified sample of students.
b. What factors would influence your choice of sample size?

The university decided finally to take a random sample of 25 second-year management students. The mean and standard deviation of the income earned by these students was $3270 and $63.10, respectively.

c. Calculate the margin of error for the mean summer income, with 99% confidence.
d. What does this margin of error mean?
e. Does this margin of error apply to all management students, or to all university students? Explain.

12. Ron Jones, the general manager of the National Paper Company (NPC), wants to determine the mean diameter of pine trees on land that is being considered for purchase. A random sample of 41 trees gives a mean of 41.42 cm with a standard deviation of 6.35 cm.
a. Why do you think the mean diameter would be of interest to NPC?
b. Construct a 95% margin of error for the mean diameter.
c. Ron thought maybe he should have constructed a 99% margin of error. Explain briefly the effects of increasing the confidence level.
d. Ron wishes to estimate the mean diameter to within 1 cm with 95% confidence. How many trees should be included in his sample?

13. A retail company hired an auditor, Francis Small, to verify the accuracy of its new invoice system. Francis randomly selected 70 invoices produced since the system was installed. She then compared each invoice against the relevant internal records to determine by how much the invoice was in error. She found that on average the error was $6.35, with a standard deviation of $35.30.
a. Identify the population Francis was studying.
b. She first calculates a 95% margin of error for the mean error per invoice. What value did she obtain?
c. She concludes from this calculation that the margin of error did not make sense, as it was larger than the average error. Is her conclusion correct? Explain.
d. Comment on the accuracy of the new invoicing system.

Chapter 8

Regression Analysis

8.1 Introduction

In this chapter and the next, we consider a variety of situations where valuable information can be obtained by investigating relationships between variables. This is the topic of *regression analysis*. Regression is widely used in business, but is also widely misused. It is very easy to become so involved with the complexities of the various regression formulae, and with endless calculations, that the purpose and scope of regression is overlooked. As usual we shall leave the calculations to Excel and focus instead on what regression means, the basic ideas involved in carrying out a regression analysis, checking whether the underlying assumptions have been satisfied, and how to use the results.

We start by considering five data sets and asking you some questions about them. This should give you some insight into regression, and what should be done next.

 Ice Cream Sales

Wiremu Harvey, the marketing manager of the New Top Ice Cream Company, is reviewing the company's sales data. The sales measured over 30 four-week periods, together with the mean daily temperature in each period, are given in Figure 8.1.

Q1. How might Wiremu use these data?

Q2. What questions should he be asking?

Q3. What should he do next?

Period	Sales (000 litres)	Mean temp (°C)	Period	Sales (000 litres)	Mean temp (°C)	Period	Sales (000 litres)	Mean temp (°C)
1	438	5	11	325	-2	21	362	7
2	425	13	12	339	-3	22	349	4
3	446	17	13	374	0	23	323	0
4	483	20	14	361	4	24	370	-3
5	461	21	15	433	13	25	351	-2
6	391	18	16	433	17	26	408	1
7	371	16	17	534	22	27	427	5
8	327	8	18	503	22	28	473	11
9	306	0	19	438	19	29	496	18
10	291	-4	20	389	16	30	523	22

Figure 8.1   Ice Cream Sales and the Mean Daily Temperature

 ## *Raw Material Moisture Content*

During a certain production process, it is necessary to decrease the moisture content of some intermediate product. A cause and effect diagram has revealed that one potential problem is the moisture content in the raw material. Figure 8.2 gives 50 pairs of values recorded on the percentage moisture content of the raw material (RM) and the moisture content of the intermediate product (PROD) made from it. (Adapted from Ishikawa's *Guide to Quality Control*; Asian Productivity Organization, 1982.)

No.	RM	PROD	No.	RM	PROD	No.	RM	PROD	No.	RM	PROD	No.	RM	PROD
1	1.10	1.40	11	1.50	1.40	21	1.30	1.50	31	1.85	2.10	41	1.05	1.85
2	1.25	1.70	12	1.40	1.50	22	1.45	1.55	32	1.40	2.00	42	1.35	2.10
3	1.05	1.30	13	1.35	1.70	23	1.20	1.55	33	1.60	2.30	43	1.60	2.10
4	1.75	1.75	14	1.20	1.40	24	1.90	1.90	34	1.10	1.60	44	1.45	2.20
5	1.30	1.30	15	1.00	1.35	25	1.65	1.70	35	1.60	1.75	45	1.20	1.80
6	1.15	1.20	16	1.40	1.30	26	1.05	1.85	36	1.85	2.40	46	1.35	1.80
7	1.70	1.40	17	1.60	1.60	27	1.60	2.05	37	1.70	2.30	47	1.05	1.70
8	1.60	1.95	18	1.50	1.85	28	1.55	2.30	38	1.55	1.90	48	1.30	2.30
9	1.50	1.50	19	1.80	1.70	29	1.40	2.00	39	1.45	2.15	49	1.45	1.80
10	1.80	1.90	20	1.65	1.55	30	1.30	1.90	40	1.15	2.00	50	1.30	1.70

Figure 8.2   Raw Material and Product Moisture Content

Q4. *What question might the production manager be interested in asking? What should she do next?*

Inflation and Interest Rates

The data in Figure 8.3 give the average annual rates of interest and inflation between 1979 and 1992.

Year	Inflation rate (%)	Interest rate (%)	Year	Inflation rate (%)	Interest rate (%)
1979	6.9	17.2	1986	3.9	9.0
1980	6.5	16.4	1987	2.5	8.3
1981	6.3	14.5	1988	3.1	7.7
1982	6.1	12.5	1989	2.4	7.5
1983	5.5	11.3	1990	1.9	6.9
1984	5.0	10.2	1991	1.3	6.2
1985	4.3	8.7	1992	0.8	6.0

Figure 8.3   Inflation and Interest Rates

Q5. If you are interested in predicting the inflation rate if interest rates were raised to 8%, what would you do next?

 A Salary Survey

In a salary survey of its members, the National Society of Accountants collects data on salary ($000), age, and qualifications. Each member's qualifications are converted into a score (QS), where the higher the score the greater the qualifications possessed. Figure 8.4 shows a sample of data from 30 members.

Q6. How do you think the Society might use this data? What questions might be of interest? What should they do next?

Salary	Age	QS	Salary	Age	QS	Salary	Age	QS
372	68	14	164	49	19	219	42	25
206	58	17	113	45	19	186	33	24
154	52	17	82	34	20	155	31	25
175	47	14	32	32	22	114	24	27
136	42	17	228	57	20	341	62	24
112	42	17	196	45	23	340	54	25
55	32	18	128	38	21	283	48	25
45	59	19	97	30	24	267	39	26
221	60	20	64	25	28	215	32	27
166	53	21	249	54	24	148	27	27

Figure 8.4   Salary Data from National Society of Accountants

   ## Noise Levels at London Gatwick Airport

A study was conducted at London Gatwick Airport to investigate the existing procedures for the prediction of aircraft noise. The aim was to predict the perceived noise level (PNL) given the slant distance in metres (SD), which is the distance from the point at which the aircraft starts its takeoff to its position when it passes over the noise recorder located one kilometre beyond the end of the runway. The data collected are shown in Figure 8.5.

No.	SD	PNL	No.	SD	PNL	No.	SD	PNL	No.	SD	PNL	No.	SD	PNL
1	993	107	13	982	99	25	246	114	37	195	127	49	1000	98
2	1019	98	14	248	117	26	192	117	38	213	117	50	342	116
3	977	102	15	204	120	27	1037	97	39	37	119	51	476	115
4	182	120	16	149	120	28	169	124	40	400	115	52	1029	100
5	295	114	17	207	116	29	36	131	41	298	120	53	1083	99
6	96	123	18	211	116	30	86	128	42	455	112	54	293	121
7	93	121	19	1037	100	31	250	118	43	328	117	55	214	123
8	994	100	20	178	115	32	375	122	44	1014	104	56	1008	98
9	136	121	21	207	115	33	1046	102	45	1072	107	57	996	99
10	204	119	22	248	111	34	1007	97	46	1024	106			
11	1015	97	23	1008	91	35	991	100	47	328	117			
12	996	101	24	205	118	36	250	121	48	234	120			

Figure 8.5   Noise Levels at Gatwick Airport

Q7. *Look carefully at the data. What, if anything, do you notice? What would you do next?*

8.2 Scatterplots

The first thing to do when studying the relationship between two quantitative variables is to plot the data. **Failure to plot the data often leads to incorrect analysis and inappropriate predictions**. A scatterplot is used to plot such data, with one variable plotted on the vertical axis (the *y*-axis) and the other variable on the horizontal axis (the *x*-axis).

A scatterplot is easily produced in Excel using the chart wizard as explained in Section A.9 in Appendix A. First the data are entered into two columns of a spreadsheet, and then highlighted. At step 1 of the chart wizard, select the *XY (Scatter)* option, and choose the type of scatterplot required. It is usual to have just the data points on the graph, without any joining lines. Finally, the graph can be customised by including axes and graph titles, and modifying the axis scales if necessary.

The scatterplot of the ice cream sales data in Figure 8.1 is given in Figure 8.6.

Q8. *How would you describe the relationship between sales and temperature?*

Q9. *How would you use this plot to predict sales from an estimated mean daily temperature? What do you estimate sales would be when the mean daily temperature is 10°C? When it is 20°C?*

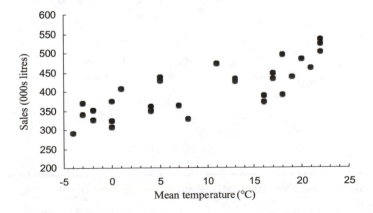

Figure 8.6   New Top Ice Cream Sales

That companies are using data, such as that in Figure 8.6, on daily temperature to help predict sales can be seen from the following newspaper extract:

> British companies are tuning in to weather forecasts to help predict consumer trends—a move that is saving them millions of pounds a year ... We are analysing historical sales statistics ... and identifying the weather factors which influenced those sales ... Mathematical equations are then devised to pinpoint the precise relationship between weather and sales to enable retailers to predict demand much more accurately. This method ... to predict fluctuations in sales of a single brand such as Unilever's Persil detergent (people do more washing when the weather is fine) or Nestle's Kit-Kat chocolate bars, which can go soft in hot weather ... The weather also has a big impact on the purchase of toilet rolls and can account for monthly variation in sales of up to 20,000 packs.[1]

Scatterplot of Moisture Content Data

Figure 8.7 shows a scatterplot of the moisture content data in Figure 8.2.

Q10. *What can you say about the relationship between the percent moisture contents of the intermediate product and the raw material?*

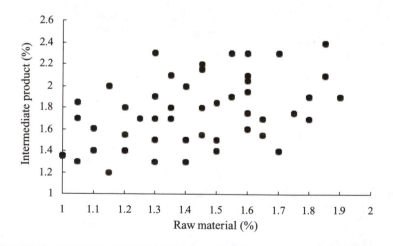

Figure 8.7   Product and Raw Material Moisture Content

---

[1]   Source: *New Zealand Herald* (Reuters), 5 August 1997.

Scatterplot of Interest and Inflation Rates

A scatterplot of the data in Figure 8.3 is shown in Figure 8.8.

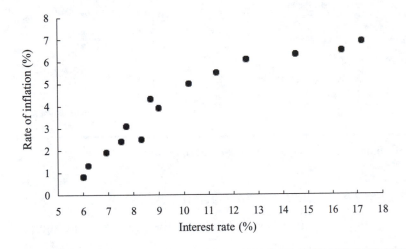

Figure 8.8   Interest and Inflation Rates, 1979–1992

> **Q11.** *How would you describe the relationship between interest rates and inflation rates over the period 1979 to 1992?*

 ## Scatterplots of Salary Survey Data

Two scatterplots of the salary survey data in Figure 8.4 are given in Figure 8.9. The first plots salary against age, and the second salary against qualification score.

> **Q12.** *What would you conclude from these plots? What other plots might you wish to see?*

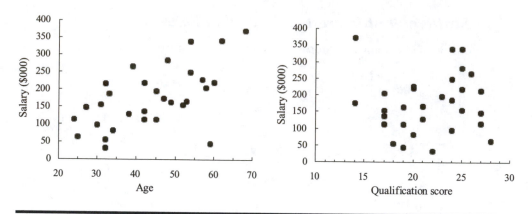

Figure 8.9   Salary Against Age and Qualification Score

Noise Levels at London Gatwick Airport

The data shown in Figure 8.5 are plotted on the scatterplot in Figure 8.10.

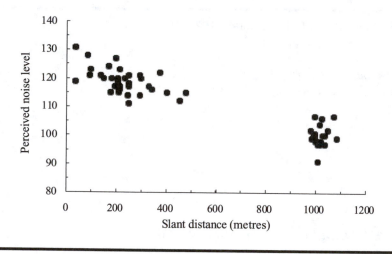

Figure 8.10   Gatwick Airport Data

Q13. *What PNL value would you predict for a slant distance of 600 metres?*
Q14. *What conclusions do you draw from this plot?*

If these two sets of points, corresponding to low and high slant distances, are plotted individually, the scatterplots in Figure 8.11 are obtained.

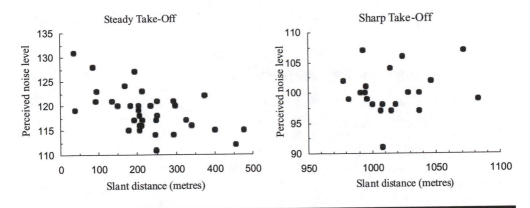

Figure 8.11   Scatterplots for Steady and Sharp Takeoff

Q15. What do you now conclude?

This example illustrates very clearly the potential dangers of trying to predict without first plotting the data. In fact, such a comment is true for any form of analysis.

Q16. What do you think is happening in this example?

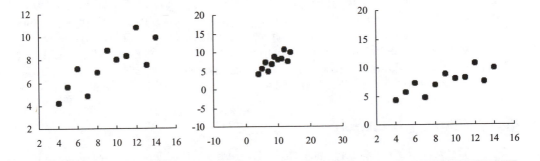

Figure 8.12   Effect of Changing Scale on a Scatterplot

Care has to be taken when using scatterplots to examine relationships between variables. More space around the data can make a relationship look stronger. Changing the spread of one variable can change the perceived strength of the relationship. All three scatterplots in Figure 8.12 have been obtained from the same set of data, but you might think that one relationship looks stronger than the others.

8.3   Relationships Between Variables

We have looked at five examples where we examined the relationship between two quantitative variables. In the ice cream example the marketing manager may be interested in trying to predict sales from mean daily temperature. This could be of considerable value if reliable weather forecasts are available. Of course, temperature is unlikely to be the only variable that could be used to predict ice cream sales. We shall consider later how two or more variables might be used to predict sales.

In the moisture content example, if there is a strong relationship between the moisture contents of the raw material and the intermediate product, then it may be possible to control the moisture content of the intermediate product by ensuring that the moisture content of the raw material is kept at certain values. For example, in New Zealand sawmills, radiata pine is dried in kilns. The moisture content of the dried wood is a critical factor in further processing, and is known to be strongly related to the moisture content of the timber before it goes into the kiln. This type of relationship can be depicted as in Figure 8.13 and can often be exploited to control the outputs.

Figure 8.13   Relationship Between Inputs and Outputs

Another situation where knowledge of the relationship between two variables can be very valuable is when the variable of interest either takes too long to measure, or is expensive to measure. If a substitute variable can be found that is strongly related to the variable of interest, then this substitute variable can be used instead. For example, the amount of shrinkage in an injection moulded plastic product is usually critical. But it takes a few hours until the product has cooled sufficiently before a final measure of shrinkage can be obtained, by which time it will be too late to make any adjustments to the injection-moulding machine. However, if the relationship between final shrinkage and shrinkage at the time of production is known, then an immediate estimate of shrinkage can be obtained.

8.4   Regression Analysis

Regression analysis is widely used in management for determining the precise form of the relationship between two or more related variables. It is also, however, one of the most misused statistical techniques. Our main aim in this chapter is to acquaint you with the role, scope, and interpretation of regression. It is more important that you understand when, and when not, to use regression than the technical details of calculating regression equations.

First consider the simple linear regression model. It involves two quantitative variables: the *response variable* and the *explanatory variable*. One of the primary purposes of regression analysis is to predict the value of the response variable for a given value of the explanatory variable. In a scatterplot it is usual to put the response variable on the vertical or *y*-axis and the explanatory variable on the horizontal or *x*-axis. For this reason the response variable is sometimes referred to as the *y-variable* and the explanatory variable as the *x-variable*. In many books the response variable is called the *dependent* variable and the explanatory variable the *independent* variable.

One common mistake in regression analysis is to confuse the two types of variables. Sometimes either could be the response variable; in other cases only one of them can sensibly be used. The question to keep in mind is "Which variable are we interested in predicting, given a value of the other variable?" For example, if the two variables are the heights and weights of a group of primary school children, then either variable could be the response variable, and which we choose depends on the purpose of the study. On the other hand, in the ice cream sales example, the response variable is clearly the sales data, as predicting daily temperature given sales of ice cream does not make any sense.

Many books do not make it clear that when they talk about regression analysis they really mean *simple linear* regression. That is, they are concerned with fitting *straight lines* to data. This may be totally inappropriate for a number of reasons, but principally when the evidence of the scatterplot suggests that the relationship is nonlinear, as in Figure 8.8. Failure to first plot and examine the data probably leads to the most frequent misuse of regression analysis.

Regression Analysis of Ice Cream Data

Consider again the ice cream sales data in Figure 8.1. The starting point of any analysis is the scatterplot, which is reproduced in Figure 8.14.

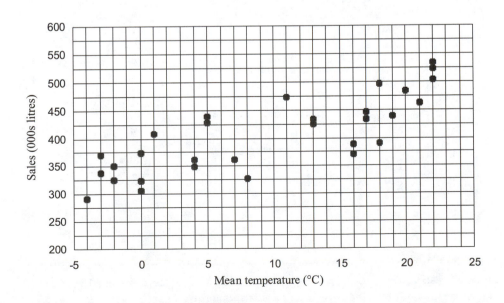

Figure 8.14   New Top Ice Cream Sales

To illustrate regression analysis we will first develop a *subjective* relationship between ice cream sales and mean daily temperature.

1. Draw a straight line on the plot, which you think best "fits" these data points.
2. Read off the value of ice cream sales where your line crosses the vertical axis **at a temperature of 0°C**; we shall denote this *intercept* by *a*. Hence, $a =$ _____.
3. Calculate the slope of the line. That is, calculate from your line the change in sales resulting from a 1°C change in temperature. To do this, read off the sales at 20°C, work out the change in sales from 0 to 20°C, and divide this by 20. The *slope* will be denoted by *b*. Hence, $b =$ _____.
4. The equation of your straight line is $y = a + bx =$ _____, where *y* is the response variable (sales) and *x* is the explanatory variable (mean temperature).

8.5 Least Squares Method

In practice, fitting a model by eye is too imprecise and can lead to substantial differences from one person to another. We need a precise criterion for determining objectively the equation of the *line of best fit* for any set of data. This is the *least squares* criterion.

Again we shall illustrate the procedure with the ice cream sales data. The scatterplot is reproduced in Figure 8.15. We have also drawn in a straight line that best fits the data according to the criterion we now describe.

We need to estimate the intercept and slope of the line so that the model fits these data well. As you will see in the next section, the least squares line of best fit to the data has the equation

$$y = 348.9 + 5.9x$$

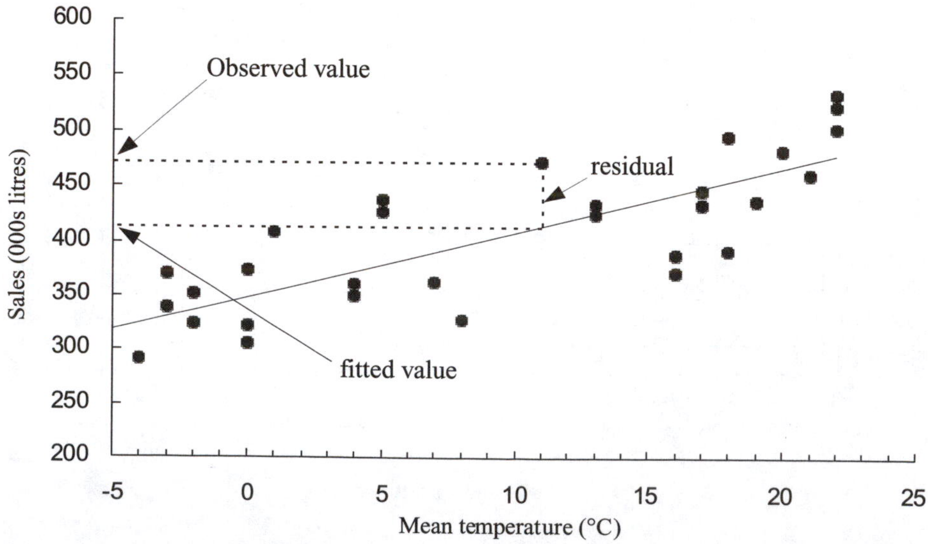

Figure 8.15   New Top Ice Cream Sales

For the value $x = 11°C$, the ice cream sales predicted by the regression line is $348.9 + 5.9 \times 11 = 413.8$; this is called the *fitted value* of y. The discrepancy between what we observed and the fitted value is called the *residual* of y (or the *error* in the fitted value). The fitted value and residual corresponding to the observed value at 11°C are shown in Figure 8.15.

Thus residuals are defined as

$$\text{Residual } (e) = \text{observed value of } y - \text{fitted value of } y$$
$$= y - (a + bx)$$

Any observed value of y is, therefore, represented by a *model* that may be written symbolically as

$$y = a + bx + e$$

which expresses the observed value as a "discrepancy" (e) from a linear trend with intercept a and slope b. To fit a regression line by least squares, we choose a and b to minimise the sum of squares of these residuals.

Apart from being the basis for least squares analysis, residuals provide important information for examining both the adequacy of the fitted model and the assumptions underlying our analysis. The study of residuals is a vital part of any regression analysis; we shall look more closely at them, and the assumptions we are making, in Sections 8.7 and 8.8.

✎ Calculating the Regression Line Using Excel

In most statistics books you will find formulae for calculating the least squares slope and intercept, but it is tedious and time consuming to do by hand. We shall use Excel instead. First we input the data into a spreadsheet. In the Excel spreadsheet in Figure 8.16, we have put the ice cream sales data (y) into cells A1:A30 and the mean daily temperature (x) into cells B1:B30. Only the first 18 values of the 2 variables are shown here.

The *Paste Function* command can be used to calculate both the intercept (a) and slope (b) by choosing *Statistical* in the function category and selecting either *INTERCEPT*, as shown in Figure 8.16, or *SLOPE* in the function name. This opens a dialogue window where the ranges containing both the x and y values are specified, and from which the intercept or slope are calculated. Figure 8.17 shows the *INTERCEPT* window; the equivalent *SLOPE* window is effectively identical.

Q17. *For the ice cream data, the intercept is calculated as 348.9, as shown in Figure 8.17. What does this intercept of 348.9 mean?*

Q18. *In a similar way, using the SLOPE function, the value of the slope is given as 5.9. What does this value mean?*

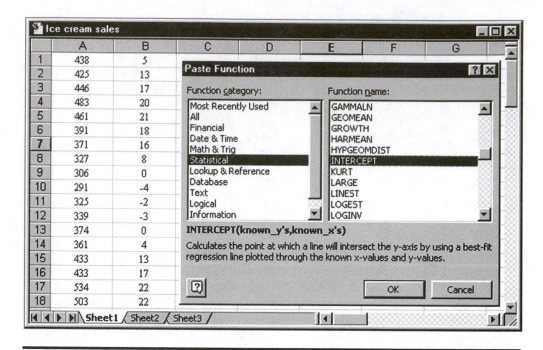

Figure 8.16   Excel Spreadsheet Windows for Calculating the Intercept

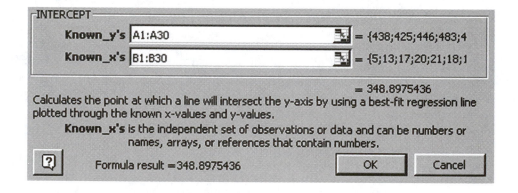

Figure 8.17   *INTERCEPT* Dialogue Window

The fitted regression equation is therefore

$$y = 348.9 + 5.9x \quad \text{or} \quad \text{sales} = 348.9 + 5.9 \times \text{mean daily temperature}$$

Q19. Given this regression line, what would be your predicted sales for a mean temperature of 10°C?

✍ *The Regression Add-In in Excel*

We have already seen how to calculate the regression line in Excel using functions in the *Statistical* function category. An alternative method is to use the regression facilities in the *Data Analysis Tools Add-In*. Choosing *Regression* from the *Data Analysis* menu gives the dialogue box in Figure 8.18.

The data ranges for *x* and *y* are entered into the appropriate *Input* boxes. If our spreadsheet contains names for the *x* and *y* variables at the head of each range, then these cells could be included in the data ranges and the *Labels* box checked, in which case the output will refer to the variables by name. None of the other dialogue options are required at this stage, although we shall later use some of the other options available. Part of the output from this analysis is given in Figure 8.19, which confirms our previous calculations for the intercept and slope.

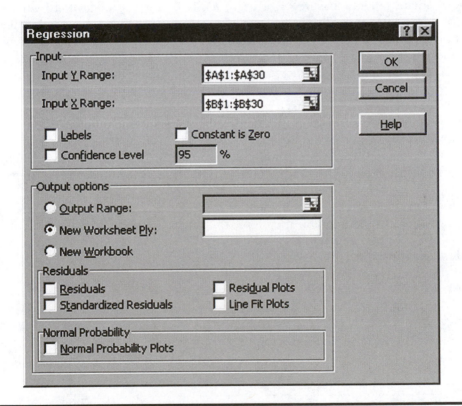

Figure 8.18 Regression Data Analysis Dialogue Window

Regression Statistics			Coefficients
Multiple R	0.8036241	Intercept	348.8975436
R Square	0.6458117	X Variable 1	5.9055217
Standard Error	40.3337604		
Observations	30		

Figure 8.19 Selected Regression Output

8.6   Correlation

The line $y = 348.9 + 5.9x$ has been fitted to the data using regression. Is it a good fit? For a perfect fit the line would go through all the data points and all the residuals would be equal to zero. However, in most practical situations the line looks like the one in the ice cream example in Figure 8.15 where most of the points do not lie on the regressions line, that is, the residuals are mainly nonzero.

We can measure the strength of the linear relationship between ice cream sales and temperature by the sum of the squares of the residuals (since the sum of the residuals is always zero). The larger this sum of squares, the worse the fit. This measure suffers from one major drawback, namely that it depends on the units of measurement of the response variable. For instance, if we were to measure ice cream sales in litres rather than thousands of litres, we would get a different sum of squares. The measure of fit used, R^2, is a measure of fit that overcomes the problem by standardising the sum of squares so it is between 0 (no linear relationship) and 1 (a perfect linear relationship) for any units of measurement. For the ice cream data $R^2 = 0.6458$, given as *R square* in Figure 8.19. This says that 64.58% of the variation in ice cream sales is explained by the linear regression model. An $R^2 = 1$ would mean that 100% of the variation is explained by the model. In other words the line would go through all the data points.

Closely related to R^2 is the *correlation coefficient r*, which measures the linear relationship between two variables. *Multiple R* given in Figure 8.19 is the square root of *R square*. If there is a positive linear relationship between two variables (positive slope) then $r = +R$, and if there is a negative linear relationship (negative slope), $r = -R$. For the ice cream data, $r = +0.8036$.

However, the correlation coefficient only measures *linear* association between two variables. It is important, therefore, to always construct a scatterplot of the data. If you observe a linear relationship then r is a good summary of the strength of that relationship. If the relationship is nonlinear, then it should be apparent from the scatterplot.

We can also calculate the correlation coefficient in Excel using the *Statistical* function *CORREL*. For the ice cream data, with sales and mean temperature data in cells A1:A30 and B1:B30, respectively, we get a correlation coefficient of 0.8036, as before. The dialogue window for *CORREL* is shown in Figure 8.20.

Figure 8.20   Dialogue Window for the *CORREL* Function

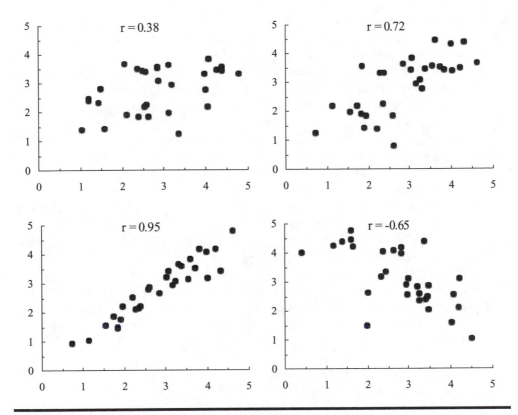

Figure 8.21 Scatterplots and the Correlation Coefficient

✺ *Strength and Sign of Correlation*

The scatterplots in Figure 8.21 give you some idea of the strength (strong or weak) and sign (positive or negative) of the correlation coefficient for different scatterplots.

For the ice cream data in Figure 8.1, the correlation coefficient between sales and mean temperature is +0.804, indicating a strong positive linear association between these two variables.

For the moisture content data in Figure 8.2, the correlation coefficient between the moisture contents of the intermediate product and raw material is +0.397.

The correlation coefficient between inflation rates and interest rates for the data in Figure 8.3 is +0.933. Although this indicates a strong linear relationship between the two variables, we can see from the scatterplot that a curvilinear relationship is even stronger, as we will see in Chapter 9.

For the Gatwick Airport example in Figure 8.11, the correlation coefficient between slant distance and noise level is −0.564 for the steady takeoff and +0.145 for the sharp takeoff. In the latter case there is little evidence of any linear relationship between the two variables, which is confirmed by the scatterplot.

8.7   Regression Assumptions

From the fitted regression model we can predict the value of the response variable *y* given a value of the explanatory variable *x*.

For example, you can predict what ice cream sales will be for a given mean daily temperature. For instance, at a temperature of 0°C you should get sales of about 350 (i.e., 350,000 litres). Of course few, if any, of the actual values fall on the line you have drawn. Looking at the data we see, for instance, that we have 3 readings at 0°C, namely 306, 374, and 323. This is quite a bit of variation. If we had a lot more data we might have other readings at 0°C. With masses of data we could build up a distribution of sales figures at 0°C. The value we read off from our regression line is the mean value of this distribution. The same applies to other temperature values.

So we can picture a distribution of sales values at any level of temperature (x), whose mean value is given by the fitted regression model. What about the amount of variation in these distributions? When we carry out a regression analysis we make the following important assumption:

Assumption 1: The standard deviation of the response variable is the same at any value of the explanatory variable. That is, the spread of the distribution is the same for all values of x. Only the mean value changes.

The validity of this assumption can be checked, as we shall see later.

Let us denote the *standard deviation* of these distributions by σ. The value of σ can be estimated from the sample data, and is part of the standard Excel regression output. It is shown in Figure 8.19 as the *Standard Error*, which for the ice cream data is 40.3. If we assume that the variation about the line of best fit is normally distributed, then approximately 95% of the data values should fall within a *data band* of width $\pm 2\sigma$ either side of the line of best fit. For the ice cream data, this band is at $\pm 2 \times 40.3 = \pm 80.6$ and is shown in Figure 8.22.

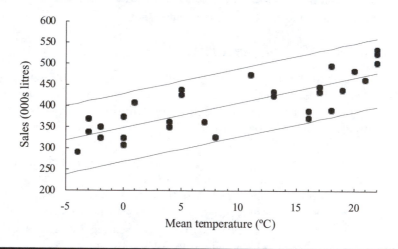

Figure 8.22   Data Band for New Top Ice Cream Sales

Another important assumption made in carrying out a regression analysis is the following:

Assumption 2: The response values are *independent* of each other. This is why regression is not a good forecasting method for data that is plotted sequentially over time, because in that case the response in one period is often related to the responses that occurred in previous periods. *Time series* methods are available which take into account these dependencies, some of which will be discussed in Chapter 10.

8.8 Analysis of Residuals

To check whether we have fitted an appropriate model, or whether our assumptions are reasonable, we should both plot the data and examine the residuals from the analysis. A scatterplot of the data should be examined before any analysis is carried out. Often this will tell you whether straight-line regression is likely to give a good fit to the data. An analysis of residuals can give information not only about the adequacy of the model but also about the assumption that each x value has a symmetric distribution with the same standard deviation σ.

A simple way of examining the residuals is to draw a scatterplot of the residuals (on the y-axis) against the fitted values (on the x-axis). This should show an even scatter about the x-axis, with approximately equal numbers of positive and negative residuals. Such a plot for the ice cream sales data is given in Figure 8.23.

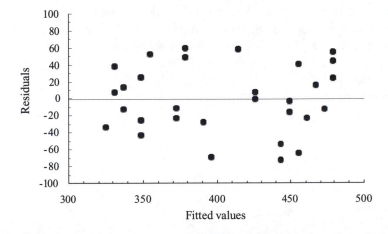

Figure 8.23   New Top Ice Cream Sales—Residual Analysis

Q20. What does this plot tell you about the linear model we have fitted?

✍ *Calculating Residuals*

A residual is simply the difference between any observed value of y, and the corresponding fitted value determined from the least squares regression line. For example, the first period in the ice cream data had observed sales of 438 when the mean temperature was 5°C, so that

$$\text{fitted value} = 348.9 + 5.9 \times 5 = 378.4$$

and

$$\text{residual} = 438 - 378.4 = 59.6$$

Q21. *What are the fitted values and residuals for the second and third periods?*
Q22. *Identify the first three points in Figure 8.23.*

In practice, we can use the *Data Analysis Regression* facility in Excel to give us the residuals, and a residual plot. For the ice cream data the regression dialogue window is shown in Figure 8.18. To obtain the residuals in addition to the standard regression output, we simply check the *Residuals* box. This produces a list of residuals, part of which is reproduced in Figure 8.24. Note that the *fitted value of y* is here referred to as the *Predicted Y*.

From these values we can obtain the plot of residuals against predicted sales, as given in Figure 8.23.

RESIDUAL OUTPUT

Observation	Predicted Y	Residuals
1	378.43	59.57
2	425.67	-0.67
3	449.29	-3.29
4	467.01	15.99
5	472.91	-11.91
6	455.20	-64.20
7	443.39	-72.39
8	396.14	-69.14
9	348.90	-42.90
10	325.28	-34.28
⋮	⋮	⋮

Figure 8.24   Sample Residual Output for the Ice Cream Data

☞ *Examining the Residual Plot*

A clear pattern in a plot of the residuals against the fitted values suggests:

- that the model is inappropriate, or
- the assumption of a constant standard deviation does not hold. For instance, a funnel effect with greater variation in the residuals as the fitted value increases suggests that larger values of *y* have greater variation.

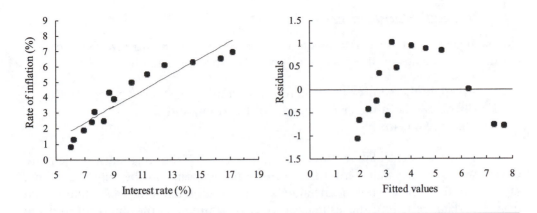

Figure 8.25   Scatterplot and Residual Plot for Inflation Rate Data

Fitting a different model to the data, or using log (y), or the square root of y, may overcome these problems.

As an example, consider the inflation data in Figure 8.3. The scatterplot of the response and explanatory variables, and a scatterplot of the residuals are given in the graphs in Figure 8.25. We can see from the scatterplot on the left that a straight line is not an appropriate model for the data. The curved pattern in the residual plot on the right confirms this. One possible solution is to extend the regression model to include a quadratic term to account for this curvature.

In a two-variable regression, the residual plot merely confirms what should be evident from the scatterplot. However, in complex situations involving more than two variables, it is difficult to draw a comprehensive scatterplot of the data and the residual plot is the simplest way of checking the suitability of the model, as will be shown in Chapter 9.

 Other Residual Plots

Plot of Residuals against the Explanatory Variable

If the model is linear in the explanatory variable, this plot should be randomly distributed about the x-axis. A curved pattern, as in Figure 8.25, suggests that the relationship is nonlinear in x. If you have several explanatory variables (see the following chapter on multiple regression), it is important to do this plot for all possible explanatory variables, whether or not you have included them in the model.

Run Chart of Residuals in Time Order

It can be useful to plot the residuals against the time order in which the data were collected. This may alert you to any relationships between observations that are close in time.

Cyclical behaviour suggests that the mean of the process is changing over time. Zigzag behaviour is often seen when there is compensatory behaviour, for example if a product is being used at a constant rate, but we ordered too much this month, and so we order less next month. Splitting, or stratifying, the data according to some other variable may help you understand the variation better.

✍ *Residual Standard Error*

The standard deviation σ is estimated by the residual standard error s_e, which is obtained following the three steps given in Section 4.5, namely:

1. calculate the sum of the squared deviations,
2. divide by the number of degrees of freedom, and
3. take the square root.

Here, we are calculating the sum of squares of the residuals, which are deviations of each y value from the corresponding point on the regression line, rather than from the overall mean value of y. The regression line is also determined from the data, requiring the estimation of *two* parameters, the intercept and the slope. Hence, *two* degrees of freedom are lost. Therefore, the standard error s_e has only $n - 2$ degrees of freedom.

Again careful examination of the residuals is important before we use the *residual* standard error s_e. If the model is inappropriate, or there are unusually large residuals, then this value will tend to overestimate σ.

8.9   Unusual Observations

Observations that do not quite belong have always been a bother for data analysts. We call such unusual observations *outliers*. It is tempting just to get rid of them. But this is a shortsighted way to proceed, equivalent to ignoring anything that violates our assumptions. We need to know the reasons behind things in order to be able to understand and then improve them. Sometimes, the outlier is an important discovery, such as a previously unknown relationship between the variables.

However, we should first check each unusual observation carefully. Is it simply a mistake? For example, typing 347 instead of 34.7. Or an impossible result, such as a machine out of order for 26 hours on Thursday. Check your data as soon as possible so that you can find out whether mistakes have been made. Correct the data where possible, and probably leave out false or incorrect numbers.

If we are satisfied it is not a mistake, then we need to ask whether this observation affects the results unduly. One way to determine this is to reanalyse the data with the unusual observation omitted. The model will probably fit better, since you took the unusual part away. But how is the model different? Are the parameters (intercept and slope) significantly different? Is there significantly less variation in the residuals?

- If it makes little difference whether the unusual observation is in or out, then leave it in.
- If it does make a difference then:
    a. you need to ask whether the model with the unusual observation omitted now appears to fit the data well, in which case you can report that the model fits for all data except this unusual one, and think why it is unusual,
    b. if a linear model still does not fit, you may need to try describing the relationship in another way.

8.10 Prediction from Regression

As we saw in Section 8.5, the regression equation can be used to predict a value of the response variable (y) for a given value of the explanatory variable (x). For example, the regression line that we obtained for the ice cream data was

$$y = 348.9 + 5.9x$$

So when the mean temperature is 10°C, the predicted sales will be 348.9 + 5.9 × 10 = 408, i.e., 408,000 litres. However, this is just a prediction based on a particular sample of 30 observations and, as with any other sample estimate, will be subject to sampling error. If we took data from another 30 periods under identical circumstances, we would almost certainly obtain a different set of results, leading to a similar (but not identical) regression line, and hence a slightly different prediction. The key question is, "How big is the margin of error associated with a regression-based prediction?"

Before we can answer this question, we need to be rather more precise about the nature of the prediction we are making. Remember what we said in Section 8.7 about the regression model and its assumptions, namely that the regression line gives the *mean (or expected) value of* the response variable for any value of the explanatory variable. In the above example, we are implicitly predicting the *average* ice cream sales on all days when the mean temperature is 10°C to be 408,000 litres. However, the average sales of ice cream at a temperature of 10°C is probably not as important to the sales manager as the likely sales in the next period when the mean temperature is expected to be 10°C. In the absence of any other relevant information, our best prediction of sales in the next period is the average sales for a mean temperature of 10°C, but there could be a much larger error associated with this figure than in a prediction of a mean value.

Q23. Why do you think this is?

In summary, any regression-based prediction could be for either:

- the *average* value of y at some value of x, or
- a *particular* value of y at some value of x.

In either case, the value we predict will be the same, namely the value of y obtained by substituting the relevant value of x into the regression equation. The difference between the two predictions is that the first one will have a smaller margin of error than the second.

8.11   Margin of Error

Most statistics books contain formulae for the margins of error for the two prediction situations we have just described. These formulae are quite complicated to work out (even within Excel!) and we will not go into the details here. Instead, we will concentrate on the general issues, so that you understand what influences the margin of error in regression situations.

✎ *Predicting a Mean Value*

As a starting point, recall the way in which we calculated the margin of error for an estimated sample mean in Section 7.6, namely

$$\text{margin of error} = \pm t_{n-1}\frac{s}{\sqrt{n}}$$

where t_{n-1} is an appropriate percentage point of the t distribution with $n - 1$ degrees of freedom, and s is the standard deviation of a random sample of n values from the relevant population. Apart from the t value (which reflects the degree of confidence in the result), the margin of error is directly proportional to the variability of the sample (s), and inversely proportional to the square root of the sample size (n).

With one minor change, this result also applies to a regression-based prediction of the mean value of y for a value of x in the *centre of the data* (to be precise, at the mean value of x). The change required is that s is the *residual standard error* (s_e) described in Section 8.8, based on $n - 2$ degrees of freedom. Consider again our prediction of the average ice cream sales when the mean temperature is 10°C, namely 408 ('000 litres). A temperature of 10°C is quite close to the mean of the x values (which is actually 9.5), and so a 95% margin of error is approximately

$$\pm t_{n-2}\frac{s_e}{\sqrt{n}} = \pm 2.048 \times \frac{40.3}{\sqrt{30}} = \pm 15.1 \quad \text{i.e., } \pm 15,100 \text{ litres}$$

> **Q24.** *How would you interpret this result?*

So what happens if our prediction is for a value of x not at the centre of the data? As you may anticipate, as we move further out from the centre of the data there is more uncertainty in our prediction. This is because any error in estimating the true slope will be amplified the further out we go, and hence a greater margin of error. It can be shown that if the x values are reasonably uniformly scattered

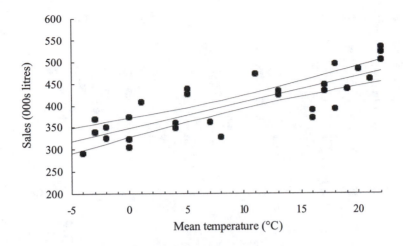

Figure 8.26   Confidence Interval for Ice Cream Sales

along the x-axis, with no obvious outliers, the margin of error at the extremes of the data will be about twice what it is in the centre.

For example, with the ice cream data, an estimate of the average sales at a mean temperature of 22°C is

$$y = 348.9 + 5.9 \times 22 = 478.7, \text{ i.e., } 478,700 \text{ litres}$$

It can be shown that this estimate has a margin of error of 26.0, i.e., 26,000 litres.[2]

Adding and subtracting the margin of error from the regression line for all values of x gives the error band, or *confidence interval*, around the regression line as shown in Figure 8.26.

✎ *Predicting a Particular Value*

In this case, we are attempting to predict the y-value of some *particular* point on the scatterplot, which will inevitably be a more uncertain process and lead to a greater margin of error. The main contribution to this margin of error comes from the natural scatter of the points around the regression line, which is measured by the residual standard error, s_e. As when predicting a mean value of y, the margin of error increases as we go out from the centre of the data, but now by a much smaller amount. In this case, the margin of error might only increase by 5 or 10% as we get to the extremes of the data. This is demonstrated in Figure 8.27, which shows the 95% error band, or *prediction interval*, for a particular period's sales of ice cream for any temperature level. If you look closely at Figure

---

[2]   The exact margin of error for a given value x is

$$\pm t \times s_e \sqrt{\frac{1}{n} + \frac{(x - \bar{x})^2}{(n-1)s_x^2}}$$

where s_x is the standard deviation of the x values.

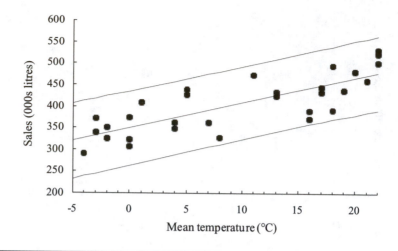

Figure 8.27   Prediction Interval for Ice Cream Sales

8.27, you can see that the error limits are slightly curved, but much less so than the corresponding band in Figure 8.26.

Compare the prediction interval in Figure 8.27 with the data band drawn at $\pm 2s_e$ in Figure 8.22, and you will see that they are effectively the same. If we make the same two predictions as before for sales at $x = 10$ and $x = 22$, the margins of error are 84.0 and 86.6, respectively, so the margin of error at the edge of the data is only 3% greater than in the centre. Furthermore, the limits of error become more parallel, and the margin of error more constant, as the size of the sample increases.

In summary, if the sample of data is reasonably large ($n > 10$), the margin of error in predicting a particular value of y is more or less constant and is approximately given by[3]

$$\text{margin of error} = \pm t_{n-2}\, s_e$$

For the ice cream data, this would be $\pm 2.048 \times 40.3 = \pm 82.6$, i.e., $\pm 82,600$ litres.

Q25. How would you interpret this figure?

---

[3] The exact margin of error for a given value x is

$$\pm t \times s_e \sqrt{1 + \frac{1}{n} + \frac{(x - \bar{x})^2}{(n-1)s_x^2}}$$

8.12 Exercises

1. Which one of the following statements about the correlation coefficient between y and x is true?
   a. it detects whether y is caused by x
   b. it provides a measure of the linear relationship between y and x
   c. it tells us by how much y increases for a unit increase in x
   d. it allows the response variable y to be predicted from explanatory variable x

2. The least squares method for fitting a regression line minimises the
   a. standard deviation
   b. residual sum of squares
   c. sum of squares of fitted values
   d. sum of absolute deviations between actual and fitted values
   e. value of the slope

3. The correlation coefficient between the age of a car and the money spent on repairs is +0.90. Which of the following statements is true?
   a. 81% of the variation in the age of the car is explained by the amount of money spent on repairs
   b. 81% of the variation in the money spent on repairs is explained by the age of the car
   c. 90% of the variation in the age of the car is explained by the amount of money spent on repairs
   d. 90% of the variation in the money spent on repairs is explained by the age of the car

4. From a regression analysis of a response variable on a single explanatory variable, a plot of residuals against fitted values is given in Figure 8.28. This plot tells you
   a. there is no relationship between the two variables
   b. the relationship is not linear
   c. the wrong explanatory variable has been used
   d. the constant variation assumption does not hold
   e. a mistake has been made in the analysis

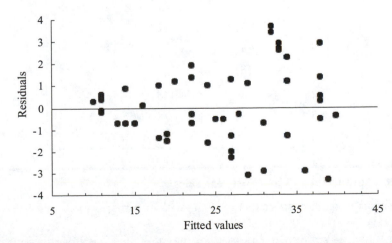

Figure 8.28   Residual Plot

5. In the manufacture of a plastic bottle top, measurements are taken on the amount of shrinkage in the top and on the speed at which the plastic is extruded. In a regression model using these two quantities, speed would be the
   a. response variable
   b. regression coefficient
   c. correlated variable
   d. explanatory variable
   e. both (c) and (d)

6. In regression a residual is defined as
   a. the horizontal distance between a point and the regression line
   b. the distance between consecutive points on the scatterplot
   c. the variation you would expect if you fitted another variable
   d. the vertical distance between a point and the regression line
   e. none of the above

7. We want to predict sales of a product from orders taken. A straight line regression is fitted, with sales as the response variable and orders as the explanatory variable. Which of the following statements is false?
   a. it is dangerous to use a regression line to predict sales when the number of orders is outside the range of values used to fit the line
   b. the intercept in the regression equation is the value of orders when sales is equal to zero
   c. a nonlinear relationship between sales and orders can be detected from a plot of residuals against fitted values

8. The personnel manager conducted an aptitude test on all sales people as they were hired. After a year's employment with the company, the manager obtained each person's total sales figures for the year. The aptitude scores and sales figures (in thousands) are given in Figure 8.29.

Score	Sales	Score	Sales
53	44	81	72
69	63	52	45
62	55	68	59
39	41	78	75
33	30	59	58
102	102	43	40
92	83	46	40
63	56	78	83
78	75		

Figure 8.29   Aptitude Scores and First Year Sales

   a. Construct a scatterplot of the data, with the response variable plotted on the vertical axis (*y*-axis).
   b. In what way might the personnel manager use the information in Figure 8.29?

c. Use Excel to fit a regression of sales on aptitude score. (Save the data for use in Question 11.)

d. Draw your regression line on the scatterplot.

e. Predict the average sales for employees with aptitude scores of 40, 65, and 90.

9. Artua Smith, a manager of a car hire firm, is trying to estimate the yearly cost of operating hire cars. He has worked out the annual costs of running a small car together with the distance travelled by each car. The graph in Figure 8.30 gives the scatterplot of his data. Artua has produced the output in Figure 8.31 from Excel.

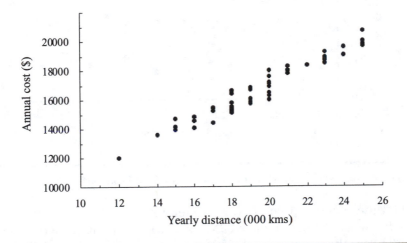

Figure 8.30   Annual Cost of Hire Cars

Regression Statistics			Coefficients
Multiple R	0.971603955	Intercept	4877.867818
R Square	0.944014246	X Variable 1	607.4185099
Standard Error	460.2415831		
Observations	50		

Figure 8.31   Excel Regression Output

a. Write down the equation of the regression line.

b. Explain the meaning of *R Square* in this regression.

c. Artua has to estimate the cost of running a car they rented out last year, which has traveled 35,000 km. What should he predict as the annual cost of this car? What 99% margin of error could he expect from this prediction?

d. What should be the annual hire out price he should charge for a similar car travelling 35,000 km next year to ensure he is 99% certain he will cover the cost plus 10% profit?

e. One of his friends, a statistician he knew at university, tells him that the hire out price could be far too low. Why?

10. In the study of noise levels at London Gatwick Airport described in Section 8.1, the scatterplot for aircraft with a steady takeoff that was given in Figure 8.11 is reproduced below as Figure 8.32. The aim is to predict the perceived noise level (PNL) from the slant distance (SD). The fitted regression equation is PNL = 124.3 − 0.024 × SD.

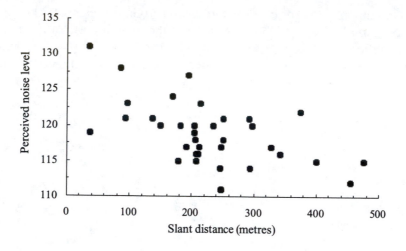

Figure 8.32   Noise Level Against Slant Distance for Steady Takeoff

    a. Plot the regression line on the scatterplot.
    b. What would be the predicted PNL value for a slant distance of 1000 metres?
    c. What are the dangers involved in predicting beyond the range of the data? Compare your predicted value with the data given in Figure 8.5.
    d. What other variables might be used to improve the prediction of PNL?

11. Consider again the aptitude and sales data given in Question 8. A plot of the residuals against the fitted values for the regression model is given in Figure 8.33.

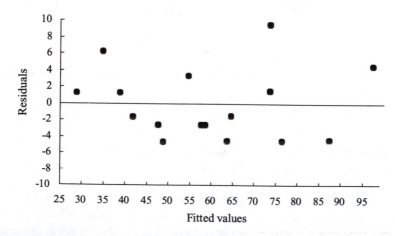

Figure 8.33   Residual Plot of Sales and Aptitude Data

a. What conclusions do you draw from this residual plot?
b. Use your spreadsheet package to calculate the standard error.
c. Calculate the 95% margin of error for the average predicted sales for employees with aptitude scores of 40, 65, and 90. Comment on your results.
d. Suppose that a prospective new employee sits for the test and obtains a score of 94. Based on the regression line, give a range of values within which you would expect her sales to be after one year. What assumptions have you made?

12. The scatterplot and the residual plot for a regression of y on x are given in Figure 8.34. One point stands out in each diagram. What do you conclude from these diagrams about the effect of this point on the regression of y on x?

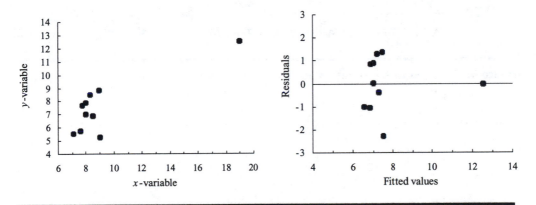

Figure 8.34   (a) Scatterplot; (b) Residual Plot

13. From Dr. G.E.J. Llewellyn and Mr. R.M. Witcomb (*The London Times*, 6 April 1977)

> Sir,
>
> Professor Mills today (April 4) uses correlation analysis in your columns to attempt to resolve the theoretical dispute over the cause(s) of inflation. He cites a correlation coefficient of 0.848 between the rate of inflation and the rate of change of "excess" money supply 2 years before.
>
> We were rather puzzled by this for we have always believed that it was Scottish dysentery that kept prices down (with a one-year lag, of course). To reassure ourselves, we calculated the correlation between the sets of figures below (see Figure 8.35).
>
> We have to inform you that the correlation is −0.868 (which is slightly more significant than that obtained by Professor Mills). Professor Mills says that, "Until ... a fallacy in the figures [can be shown], I think Mr. Rees-Mog [editor of the *Times*] I have fully established my point." By the same argument, so have we.

Comment on this letter. What conclusions do you draw?

Year	Cases of dysentery in Scotland (000s)*	Increases in prices one year later	Year
1966	4.3	2.5	1967
1967	4.5	4.7	1968
1968	3.7	5.4	1969
1969	5.3	6.4	1970
1970	3.0	9.4	1971
1971	4.1	7.1	1972
1972	3.2	9.2	1973
1973	1.6	16.1	1974
1974	1.5	24.2	1975

* Annual Abstract of Statistics, 1976, Table 68.

Figure 8.35   Cases of Dysentery in Scotland Against Price Increases

Chapter 9

Multiple Regression

9.1 Introduction

Often there will be more than one explanatory variable that could be used to predict the response variable. For instance, in the ice cream example in Chapter 8, sales may depend not only on daily temperature but also on, for example, the price of the product and the monthly advertising expenditure. Thus, we could be looking at a regression analysis that involves up to three explanatory variables. In the moisture content example we could also consider ambient temperature, processing time, the storage time of the raw material, and where it has been stored as additional explanatory variables. Some of the variables may not provide us with any additional information in predicting the response variable. We need, therefore, to look carefully at which variables we should include in our model and which should be left out. A technique for studying more than one explanatory variable is called *multiple* regression.

The first step in any multiple regression analysis is to plot the data. For the inflation rate data, it is clear from the scatterplot in Figure 8.8 that another term should be added to the model to take account of the curvature. One way of doing this is to add a second explanatory variable, which is the square of the first. Thus we are fitting a *quadratic* (multiple) regression model, and although there are now two explanatory variables (x and x^2), we still have a two-dimensional scatterplot.

With two or more unrelated explanatory variables, however, plotting the data causes some difficulties. A series of two-dimensional scatterplots can be drawn, but sometimes these can be misleading. Computer software is available to examine three-dimensional scatterplots, but this is not provided in most spreadsheet packages. As we mentioned in the previous chapter, this is why residual plots become particularly important in multiple regression.

9.2 Multiple Correlation Coefficient

In Section 8.6 we defined R^2 as a measure of the strength of the relationship between the response variable y and the explanatory variable x. More generally, it gives a measure of the relationship between y and a set of explanatory variables.

179

The quantity R is called the *multiple correlation coefficient*. The correlation coefficient r is a special case for two variables y and x. The formula for R is complicated, but most spreadsheets and packages for multiple regression give R as part of the output. We need to understand how to use this value.

In general, $100R^2\%$ of the variation in the response variable is explained by the multiple regression on the explanatory variables. Adding another explanatory variable to the regression model cannot reduce the amount of explained variability in the response variable. The important question is by how much does the value of R^2 *increase* when another explanatory variable is added. There is no point adding further terms to the model, and thereby increasing its complexity, for very little improvement in the value of R^2. Methods are available to help us decide whether the increase in R^2 is *significant* or not, but these are beyond the scope of this chapter.

9.3   Multiple Regression

To illustrate the process of multiple regression, and the use of Excel, we will consider again the salary survey of members of the National Society of Accountants. The Society would like to devise a formula that would allow accountants to predict what salary they should be earning. The Society feels there are a number of explanatory variables that could be used in such a formula. The two for which data are available are the age and qualifications of the person surveyed.

Q1. *List other variables that could be important.*

A sample of data showing the salary, age, and qualifications of 30 members was given in Figure 8.4.

As a first step we should draw three scatterplots, one for each pair of variables. The results are shown in Figure 9.1, where the first two are plots of the response variable (salary) against each of the two explanatory variables (age and qualifications). The third plot is a scatterplot of one explanatory variable (qualifications) against the other (age).

Q2. *What can you say about the relationships between the response variable and the two explanatory variables?*

Q3. *Which variable would you use to predict salary? Would you use both? Why or why not?*

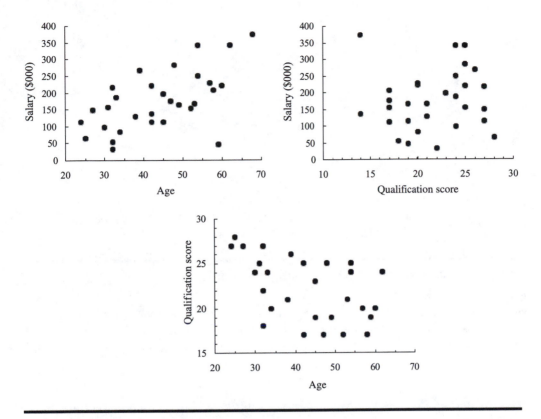

Figure 9.1 Scatterplots for the Salary Data

✎ *Regression of Salary Against Age and Qualifications*

From the scatterplots in Figure 9.1 there appears to be a linear relationship between salary and age, but not between salary and qualifications. Suppose we fit a straight-line regression of salary on age. The regression equation is

$$\text{Salary} = -15.5 + 4.36 \text{ Age}$$

Q4. How do you interpret the slope (+4.36) in this relationship?

Q5. What should an accountant be earning, on average, at age 45?

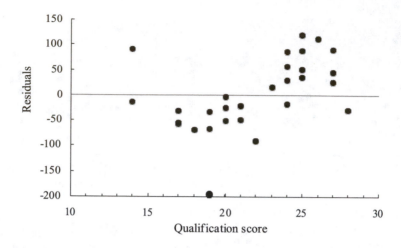

Figure 9.2   Salary Residuals Against Qualification Score

Suppose we now calculate the residuals from this line and plot these against the qualification scores to give the scatterplot in Figure 9.2. This scatterplot shows us whether there is any relationship between qualifications and the element of salary that is not age related.

Q6. When we remove the effect of age, do you think there is a clear relationship between salary and qualifications?

Q7. Is it worth including the qualification scores in our regression model? What effect will it have?

 *Multiple Regression in Excel*

Let us now fit a regression of salary against both age and qualification score. Again we estimate the intercept and the slopes of the explanatory variables using least squares, which is easily done with Excel. First enter the data on the three variables into an Excel spreadsheet, as in Figure 9.3 (only the first 21 data values are shown here). Notice that we have entered variable labels in cells A1:C1, which will subsequently appear in the regression output. Selecting *Regression* in the *Data Analysis Tools*, the regression dialogue window in Figure 9.3 is obtained. Note that in the *Input* boxes we have included the variable names (and checked the *Labels* box), and that the *X Range* box includes both explanatory variables, in cells B1:C31. We also check the *Residuals* box to produce a listing of the residuals, which will be used to obtain the residual plot in Figure 9.5.

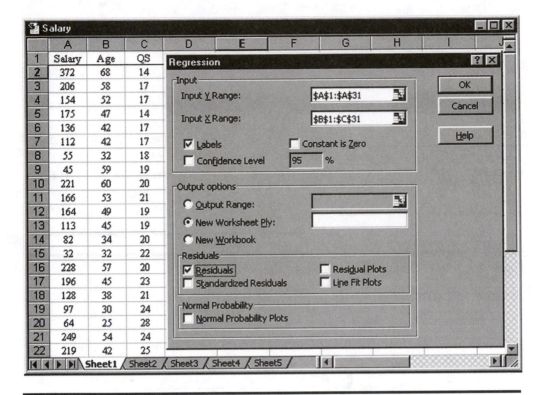

Figure 9.3   Excel Spreadsheet Windows for Fitting a Regression Model

The relevant parts from the Excel regression output are shown in Figure 9.4; only the first 10 residuals are shown. From this output, the regression equation is

$$\text{Salary} = -362.6 + 6.41 \times \text{Age} + 11.89 \times \text{QS}$$

Q8. *What would you predict an accountant should be earning, on average, at age 45, but who has a qualification score (QS) of only 15?*

In Questions 5 and 8 you have made two predictions of what an accountant aged 45 years will earn on average.

Q9. *For a 45-year-old accountant with a qualification score of 15, which do you expect to be the more accurate prediction, and why?*

SUMMARY OUTPUT

Regression Statistics	
Multiple R	0.75568
R Square	0.57106
Standard Error	59.78893
Observations	30

	Coefficients
Intercept	-362.57
Age	6.41
QS	11.89

RESIDUAL OUTPUT

Observation	Predicted Salary	Residuals
1	239.8	132.2
2	211.4	-5.4
3	172.9	-18.9
4	105.2	69.8
5	108.8	27.2
6	108.8	3.2
7	56.6	-1.6
8	241.6	-196.6
9	259.9	-38.9
10	226.9	-60.9
⋮	⋮	⋮

Figure 9.4   Excel Output for Salary Regression

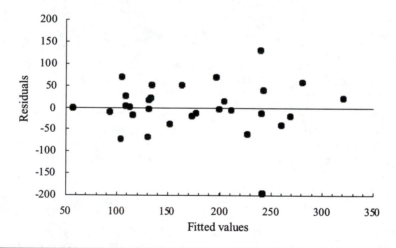

Figure 9.5   Residual Plot for Salary Against Age and Qualification Score

A plot of the residuals of this regression against its fitted values in Figure 9.5 shows no unusual patterns, although there is one individual (number 8) who seems to have an unusually low salary for his age and qualifications. This person stands out slightly on the first scatterplot in Figure 9.1, but much more clearly on the residual plots in Figures 9.2 and 9.5.

Q10. *What might have caused this unusually large residual?*

The R^2 value for the regression of salary on age alone is 36.2%. From Figure 9.4, the regression of salary on both age and qualifications gives an R^2 value of 57.1%, an improvement of over 20%. This suggests that both explanatory variables should be used in the regression model, a conclusion that was not obvious from the two-dimensional scatterplots in Figure 9.1.

9.4 Polynomial Regression

Consider again the relationship between inflation rates and interest rates for the data given in Figure 8.3. The scatterplot is reproduced in Figure 9.6.

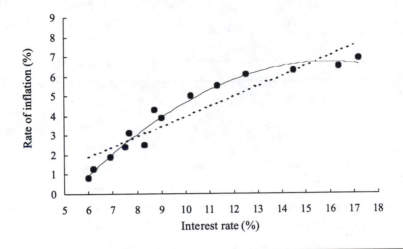

Figure 9.6   Linear and Quadratic Fits for Inflation Data

We would expect a quadratic curve to give a better fit to these data than a straight line, as can be seen from the best fitting straight line and curve on the scatterplot. The improvement we get from fitting a quadratic curve can be determined by including an additional variable in our regression to allow for the curvature. For a quadratic model we must add the square of the interest rate variable. Again the best fitting quadratic model is obtained by the method of least squares using Excel.

✍ *Fitting a Quadratic Model in Excel*

To calculate a quadratic regression model for the inflation rate data, we enter the inflation rates (y) into cells A2:A15 and the interest rates (x) into cells B2:B15, with the variable names in row 1. We now calculate the x^2 values in cells C2:C15 (i.e., C2 = B2*B2, and copy C2 to cells C3:C15) to give the entries in the spreadsheet shown in Figure 9.7. The Regression dialogue window is essentially the same as in Figure 9.3. Note that in the *Input X Range* box we have included both the x and x^2 variables.

Part of the output from fitting the quadratic model is shown in Figure 9.8. From Figure 9.8 the following quadratic regression model is obtained from Excel

$$\text{Inflation} = -8.23 + 1.88 \times \text{Interest} - 0.06 \times \text{Interest}^2$$

or

$$y = -8.23 + 1.88x - 0.06x^2.$$

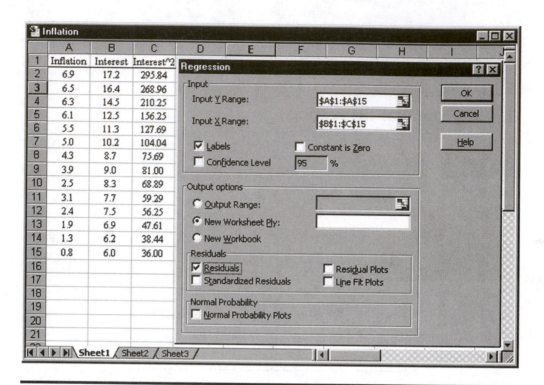

Figure 9.7   Excel Spreadsheet Windows for Fitting a Quadratic Model

SUMMARY OUTPUT

Regression Statistics	
Multiple R	0.98627
R Square	0.97273
Standard Error	0.37008
Observations	14

	Coefficients
Intercept	-8.23454
Interest	1.88459
Interest^2	-0.05939

RESIDUAL OUTPUT

Observation	Predicted Inflation	Residuals
1	6.61007	0.28993
2	6.69884	-0.19884
3	6.60499	-0.30499
4	6.04295	0.05705
5	5.47766	0.02234
6	4.80922	0.19078
7	3.66608	0.63392
8	3.91609	-0.01609
9	3.31610	-0.81610
10	2.75551	0.34449
11	2.55914	-0.15914
12	1.94152	-0.04152
13	1.16693	0.13307
14	0.93493	-0.13493

Figure 9.8   Regression Output from Quadratic Model

Q11. Do you think a quadratic model is a sensible one to use?

Q12. Use this regression equation to predict the rate of inflation given an interest rate of 19%.

Q13. *What would happen to your predicted inflation rate for a very high interest rate, such as 100%?*

Q14. *Intuitively, what would you expect your model to look like in this last situation?*

A margin of error could also be attached to any predicted value of the response variable. The *Standard Error* from Figure 9.8 will be needed for this calculation which, in multiple regression, is complicated. It will not be considered further, as specialist statistical packages are needed to give this margin of error.

✍ Analysis of Residuals

If we now plot the residuals in Figure 9.8 against the fitted values, we see from Figure 9.9 that the quadratic model seems to be adequate.

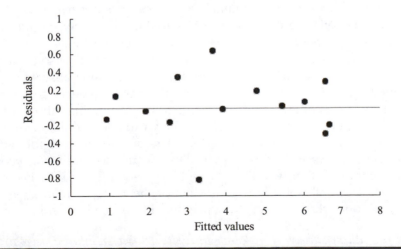

Figure 9.9   Residuals for Quadratic Model

From Figure 9.8, the multiple *R Square* value is 97.3%, which indicates that a quadratic model explains most of the variability in inflation rates. We could fit a cubic model, which would require the addition of an x^3 term, or even a higher order polynomial, but would this be worthwhile? Figure 9.10 gives the R^2 value for three different models.

Model	R^2
Linear	0.8700
Quadratic	0.9727
Cubic	0.9742

Figure 9.10 R^2 Values for Inflation Data

Q15. What do you conclude from these figures?

9.5 Using Dummy Variables

One of the key characteristics of regression analysis is that explanatory and response variables should be measurable and give interval data. In a salary survey, variables such as salary, age, qualification score, and years of work experience give interval data, and can be used in a regression analysis. However, how can we take into account attribute data relating to factors such as sex, degree classification, or type of employment, which might be important factors in explaining salary differences? We could assign arbitrary numerical values to the different categories, such as degree classification, but the results of the analysis are then dependent on the values assigned. If we assign different numerical values we will get different answers. One situation where the assignment of numerical values does not matter is when there are only two categories. In this case we can introduce a *dummy variable* that gives each alternative a numerical value (usually 0 and 1), and treat it as if it were interval data. Thus, binary variables such as sex and whether or not a person is in full-time employment can be used as dummy variables in a regression analysis.

The method can be extended to more than two categories. Suppose, for instance, that the attribute variable degree classification has three categories, namely first, second, and third class. Two dummy variables would now be needed. The first would have a score of 1 for a first class degree, and a 0 for each of the other two classes. The second dummy variable would score 1 for a second class degree and 0 for the other two. This choice for the dummy variables is not unique, but other choices will not affect the results of the multiple regression analysis. For four categories three dummy variables would be needed, and so on.

Q16. In the salary survey, what other binary variables could be used in a regression analysis?

Moisture Content Example

In Figure 8.2 there are 50 pairs of values for the percentage moisture content of some intermediate product and the moisture content of the raw material. The correlation between these two variables was not very high (+0.397), and so the raw material moisture content would not in general be a very accurate predictor of that of the intermediate product. However, it subsequently came to light that these moisture contents were measured on 2 different instruments, the first 25 pairs of values having been taken using instrument A and the last 25 pairs of values recorded using instrument B. A revised scatterplot of the 50 pairs of values, showing the instrument used, is given in Figure 9.11.

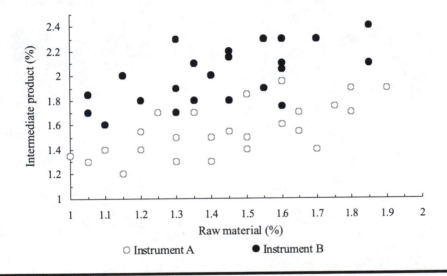

Figure 9.11   Percent Moisture Content from Two Instruments

In a regression analysis, interest would probably centre on predicting the moisture content of the intermediate product (y) given values of both the moisture content of the raw material (x_1) and the instrument used. To include the instrument in the analysis we can introduce a *dummy* variable x_2 that takes the value 0 for one of the instruments, A say, and the value 1 for the other instrument, B. Using Excel, the fitted regression model is given by

$$y = 0.72 + 0.58x_1 + 0.46x_2$$

Q17. *How would you interpret the three regression coefficients (0.72, 0.58, and 0.46) in this model?*

The effect of introducing the dummy variable x_2 is dramatic. As mentioned above, the correlation between the moisture content of the raw material and that of the intermediate product was only +0.397, giving an R^2 of 15.7%. The above model, including the x_2 variable, has an R^2 value of 71.7%, an increase of 56% in the explained variability.

As the dummy variable x_2 only takes values 0 or 1, the regression model above corresponds to two distinct parallel straight lines for instruments A ($x_2 = 0$) and B ($x_2 = 1$).

Q18. Draw these regression lines on the scatterplot in Figure 9.11, and comment on your analysis.

Q19. What difference would it make if we were to fit straight lines for instruments A and B individually?

Q20. How would you decide whether to fit a single model with a dummy variable to represent the instrument, or to fit separate models for each instrument?

9.6   Exercises

1. From a regression analysis of a response variable on a single explanatory variable a plot of residuals against fitted values is given in Figure 9.12. This plot tells you that
   a. the wrong model has been fitted
   b. there is no relationship between the two variables

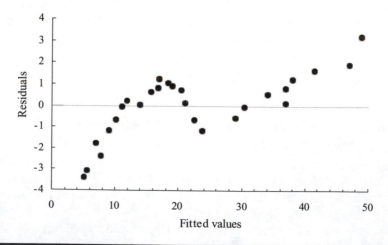

Figure 9.12   Residual Plot

c. the slope of the regression line has been wrongly calculated

d. the response variable is not normally distributed

2. A company is investigating a possible relationship between the aptitude scores prospective employees obtained in a preinterview test, the age of the applicants, and the degree obtained at university (first, second or third class). They fitted the following model to their data

$$\text{score} = 35 + 1.2 \text{ age} + 5d_1 + 3d_2$$

where d_1 is 1 for an applicant with a first class degree and 0 otherwise, and where d_2 is 1 for an applicant with a second class degree and 0 otherwise. The predicted aptitude score for a 20-year-old applicant with a third class degree is

a. 59

b. 62

c. 64

d. 67

e. not determined from the information given

3. We want to predict the annual salary of a company's executives using months of service and gender as explanatory variables. Gender is included so that we can compare salaries of female and male executives. The scatterplot in Figure 9.13 is obtained from a random sample of 40 executives from the company.

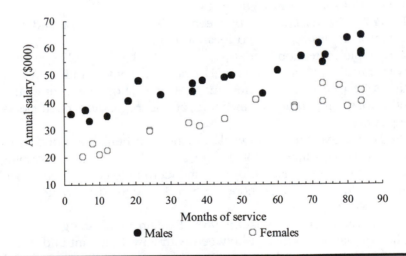

Figure 9.13   Salaries of Male and Female Executives

The following regression equation is obtained:

$$\text{salary} = 35.1 + 0.3 \text{ months} - 14.8 \text{ gender}$$

where gender = 1 if the executive is female, and gender = 0 if male.

a. What effect does months of service and gender have on annual salary?

b. Predict the annual salary expected by male and female executives after the company has employed them for 3 years.

c. Plot the regression lines on the scatterplot, and comment on the appropriateness of the model chosen.

d. Explain how you would calculate the residuals from this analysis.

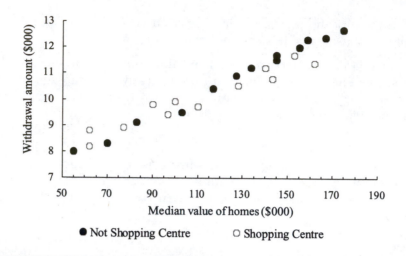

Figure 9.14   Customer Withdrawals from ATMs

4. A bank would like to be able to predict the total amount of money customers withdraw from automatic bank tellers (ATMs) on weekends based on the median value of homes in the neighbourhood and on whether the ATM is located in a shopping centre or not. A random sample of 15 ATMs is selected with the results shown in Figure 9.14.

   a. Describe the relationship between withdrawal amount and the median value of homes for the two location types.

   b. A multiple regression model is $y = 6.13 + 0.037x_1 - 0.17x_2$ where y is the withdrawal amount, x_1 the median value of homes, and x_2 the location of the ATM (where $x_2 = 1$ for an ATM at a shopping centre and $x_2 = 0$ otherwise). Interpret the meaning of the regression coefficients in this equation.

   c. Predict the average withdrawal amount for a neighbourhood in which the median home value is $130,000 for an ATM located in a shopping centre.

   d. Would it be appropriate to use the model to predict the average withdrawal amount for a neighbourhood in which the median value of homes is $300,000? Why?

   e. The R^2 value for this model is 96%. Explain its meaning.

   f. The correlation coefficient between withdrawal amount and median value of homes is 0.978. What conclusions do you draw from your analyses?

5. Figure 9.15 gives the amount of life insurance, annual income for the past year, and marital status of a random sample of 12 managers in the 45 to 50 age group.

   a. Construct a scatterplot of the amount of life insurance against annual income, with the response variable on the vertical axis (y-axis).

   b. Describe the relationship between the two variables.

   c. Label each point on the scatterplot with the marital status of the manager. What do you conclude about the impact of marital status on the amount of life insurance purchased?

Manager	Amount of life insurance ($000)	Annual income ($000)	Marital status
1	790	60	Married
2	450	85	Single
3	350	75	Single
4	1350	150	Married
5	950	80	Married
6	500	110	Single
7	250	60	Single
8	650	130	Single
9	800	160	Single
10	1400	140	Married
11	1300	120	Married
12	1200	90	Married

Figure 9.15 Life Insurance, Income, and Marital Status of Managers

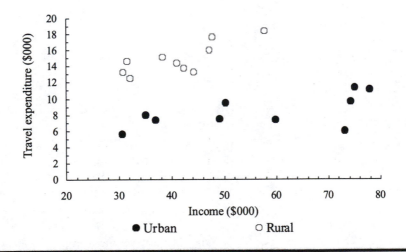

Figure 9.16 Travel Expenditure Against Income

6. The scatterplot in Figure 9.16 shows the annual expenditure on travel (y) against the family income (x) for a sample of 20 families, some of whom live in rural locations and others in urban (town/city) locations. Two possible approaches to a regression analyses of the data are being considered, namely
   1. To separate the data into two groups corresponding to rural or urban location, and to fit a regression line to each group individually.
   2. To include a dummy variable (d) to represent location (1 = rural, 0 = urban) and determine one multiple regression equation relating y to both x and d.

a. Explain which of the two regression models you would choose, and why.
b. Using both approaches, the regression equations obtained are

Model 1 $y = 7.54 + 0.178\ x$ Rural location

$y = 4.76 + 0.063\ x$ Urban location

Model 2 $y = 3.60 + 0.083\ x + 7.86\ d$

In the context of this question, explain the meaning of the two regression coefficients in the first equation of model 1, and the three regression coefficients in model 2.

c. Give two other variables that could help to explain more of the variability in a family's expenditure on travel.

Chapter 10

Forecasting

10.1 Introduction

As we have seen, one of the main reasons why decision-making is seldom straightforward is the fact that most situations involve uncertainty, which makes it difficult to anticipate future events. Indeed, many aspects of business planning and decision-making require forecasts of future circumstances and the only thing that can usually be said with confidence is that forecasts will be inaccurate to some extent. This does not mean that forecasting is a waste of time, because a plan of action based on a well-conceived forecast is likely to be better than instinctive reactions to events that occur.

As with most aspects of management, good forecasting results from a combination of technical skills (data analysis) and informed judgment. It would be foolish to completely ignore any previous data, but it would be equally short-sighted not to make use of relevant contextual information that might have a bearing on future events.

As an introduction to some of the issues that arise in forecasting, we shall consider the following three examples.

 ### *Electric Fence Insulators Inc.*

Figure 10.1 shows the monthly sales figures for an electric fence insulator for the period January 1996 through to June 1998.

Q1. *What is the main feature of these data?*
Q2. *What would you forecast sales to be in July 1998?*

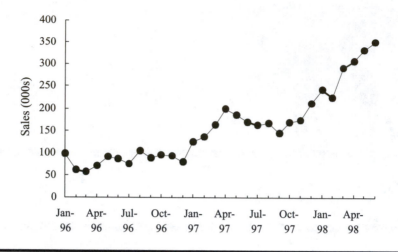

Figure 10.1   Electric Fence Insulator Sales

Plutomania

Figure 10.2 shows the daily attendances at the Plutomania theme park over a 4-week period.

> *Q3. What do you notice in particular about these data?*

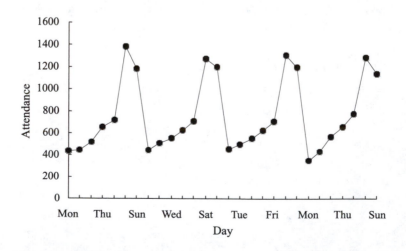

Figure 10.2   Daily Attendance at Plutomania

Pine Products

Pine Products (Australia) Ltd. manufactures a range of high quality wooden garden furniture that is sold to the domestic market. Every month, the sales director receives a new forecast of quarterly sales for the coming year.

Q4. *Why do you think regular forecasts of quarterly sales would be useful to the sales director?*

Q5. *What would be the first thing you would do to produce such forecasts?*

A run chart of quarterly sales ($mil) over the past 4 years is shown in Figure 10.3.

Q6. *What sort of things can you see from the run chart that might influence your forecasts?*

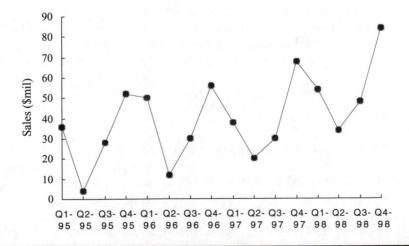

Figure 10.3 Quarterly Sales of Pine Products

10.2   Time Series

The data in Figures 10.1, 10.2, and 10.3 are measured over time at regular time periods. This type of data is called a *time series*. In our examples, we have time series of 30 monthly, 28 daily, and 16 quarterly values, respectively. Our aim is to analyse a time series in order to detect patterns that will enable us to forecast future values of the series.

The process of forecasting future values of a time series should be, in part at least, an objective one based on an analysis of relevant past values of the series. The process usually involves the following stages:

1.   Choose a forecasting model. A forecasting model is a well-defined proce-dure for calculating a forecast of future values based on the relevant past data. For example, a very simple model might be to average the last three values of the series to give a forecast of the next value.
2.   Fit the model to these past observations (that is apply it retrospectively) and obtain the *fitted values* and *residuals*. This tells us what our forecasts from the model would have been in the past (the fitted values) and how inaccurate they were (the residuals).
3.   These residuals, as in regression analysis, provide us with information about the adequacy of the chosen forecasting model (that is how well it explains the past data). If we judge the residuals to be "small" relative to alternative forecasting methods, then the model should be acceptable.
4.   If the model is deemed to be acceptable, we can use it to *forecast* future observations, such as sales in the next quarter.
5.   Finally, as new observations become available we can compare them with our forecasts, leading to an evaluation of *forecasting errors*. This allows us to *monitor* the performance of the model.

When deciding on a suitable model for a time series, it is often useful to be able to identify particular features in the past data so that the series can be broken down (decomposed) into its component parts. Given some model for the way these components fit together to form the observed values, it is possible to create forecasts by extrapolating the individual components and then reassembling them according to the specified model. It is rather like forecasting a company's total sales by predicting each product individually and then combining these to give a total sales figure.

In practice, three standard features (components) of a time series are usually defined.

■   *Long-term trend*. This is a fundamental rise or fall in the data over a sufficiently long time period. In many cases trends are assumed to be linear (and modeled accordingly), but trends can be non-linear, such as when a product's sales slow down (or even fall) after an initial period of growth.
■   *Seasonal effect*. This arises when the data are influenced by particular effects, such as the weather, that cause a regular and repeating pattern to occur over some period such as a week or a year. Many items, such as

holidays or sports goods, have sales patterns that vary predictably through the year. Likewise, the daily sales turnover in a grocery store will often follow an identifiable pattern through the week.

■ *Cyclical effect.* This is a regular, oscillating pattern that is often due to long-term economic influences, such as business cycles or the state of the world economy. It gives rise to regular, underlying swings (cycles) in the data, and as these cycles usually cover a long period they are often not analysed explicitly, but seen as part of a varying trend.

Q7. Identify which time series components occur in each of the examples in the previous section.

If we can determine which components actually exist in a time series, we can develop good forecasting models. In practice, however, this is made more difficult by the presence of a "nuisance element" in most time series, namely random variation. There are almost always a variety of unpredictable factors that serve to disrupt otherwise regular patterns and increase uncertainty. These random disturbances would normally be expected to average out to zero in the long run.

We can determine whether the time series is stable (or *stationary*), growing, or declining from period to period by using a procedure that explicitly identifies and models the trend. However, the trend may be difficult to describe by a simple model, but if some form of averaging is used to *smooth* the series, it will enable us to identify the trend more clearly from the data.

10.3 Identifying the Trend

A simple way to identify the trend component is to fit a regression model to the time series. Often a linear, or possibly a quadratic, regression will be adequate. The methods of Chapters 8 and 9 can be used, where the response variable is the quantity to be forecasted (sales, attendances, production, and so on) and the explanatory variable is time. Figure 10.4 shows the linear regression trend line for the electric fence insulator sales data.

Forecasting future sales requires extrapolation of the trend, and this can be problematic especially as you try to project further into the future. It is generally necessary to use your judgment for this extrapolation, using any additional information that is available at the time.

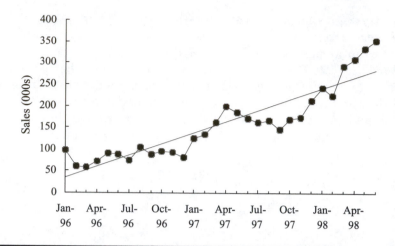

Figure 10.4   Fitting a Trend Line to the Insulator Sales Data

Q8. From the trend line in Figure 10.4 forecast sales in July 1998.
Q9. Is your forecast a sensible one? If not, why not?

It is clear from the data in Figure 10.4 that the trend has become more markedly upwards since September 1997, and this change is not adequately allowed for in the linear regression. In fact, it appears that there was little or no upward trend prior to December 1996, and this was followed by a steep upward trend over the next 5 months and then a downward trend for 6 months.

It may be appropriate, in this situation, to fit a more complex trend function. This can easily be done in Excel when editing a chart. From the *Chart* menu select *Add Trendline*. This leads to the dialogue window shown in Figure 10.5. A linear trend is the default option, but a quadratic trend could be fitted by highlighting *Polynomial* and setting *Order* equal to 2 (since squared terms are fitted in a quadratic model). As can be seen, other functions could also be tried.

Care must be exercised in using regression analysis to identify a trend. In particular, it can be dangerous to use regression analysis on data with a seasonal pattern, as the size of the seasonal effects will influence the calculation of the regression line, leading to a biased estimate of the trend. For example, consider the (artificial) data in the Excel spreadsheet in Figure 10.6. The observed values are the sum of two components. The first is a trend that starts at 20 and increases by 4 for each subsequent period. The second component is a 3-period seasonal factor of +15 in the first period, −5 in the second period, and −10 in the third period. For simplicity there is no random component. The observed values are shown in the run chart in Figure 10.6, with the trend represented by the dotted line. If the least squares regression line is calculated, this gives a value of 25 in

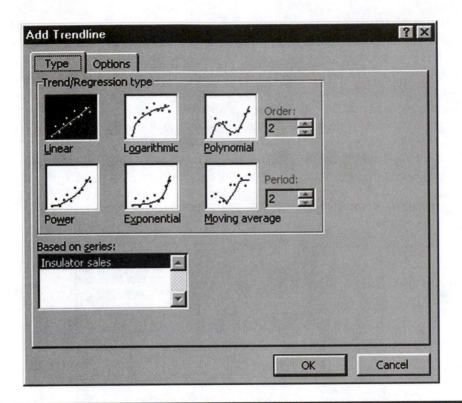

Figure 10.5   The Add Trendline Dialogue Window

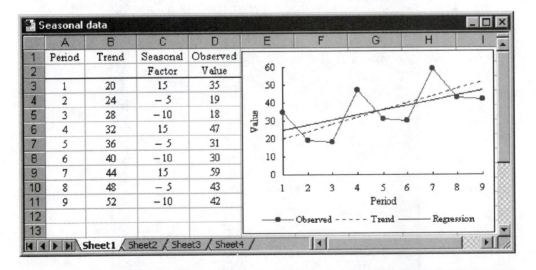

Figure 10.6   Seasonal Data with Trend

period 1 and a slope of 2.75, as shown in Figure 10.6, so that the steepness of the trend is significantly underestimated.

When the trend is difficult to describe by a simple function, or is influenced by seasonal effects, then it may be necessary to apply a smoothing technique to identify the trend. A simple, and widely applied, method is *moving averages*.

The objective of using a moving average is to remove the random variation, and, as we will see, any seasonal variations that may be present, and so reveal the underlying trend. The key question is how much smoothing to apply. With too little smoothing the random variation will not be completely "ironed out," and so may hide the real pattern. With too much smoothing, detailed features of the trend, and possibly other effects, may also be eliminated. The electric fence insulator sales data will be used again to illustrate the calculation of a moving average. These calculations are given in Figure 10.7.

The data are given in column B. Column C, headed MA(5), gives the moving averages based on 5 months. The Mar-96 value (75.2) is the average (mean) of the sales in that month (57) and the sales in the previous two months (98 and 60) and in the following two months (71 and 90). That is, it is the average of five successive values centred on March 1996. The next value (73.0) is given by

	A	B	C	D
	Month	Sales	MA(5)	MA(7)
1				
2	Jan-96	98		
3	Feb-96	60		
4	Mar-96	57	75.2	
5	Apr-96	71	73	76.9
6	May-96	90	76	77.7
7	Jun-96	87	85.4	81.9
8	Jul-96	75	89	87.3
9	Aug-96	104	90	90.4
10	Sep-96	89	91.2	89.0
11	Oct-96	95	92.2	94.3
12	Nov-96	93	96.2	102.9
13	Dec-96	80	105.4	111.1
14	Jan-97	124	118.8	127.0
15	Feb-97	135	140.2	139.9
16	Mar-97	162	161.2	150.9
17	Apr-97	200	170.4	162.7
18	May-97	185	176	168.9
19	Jun-97	170	177	170.3
20	Jul-97	163	166	171.3
21	Aug-97	167	162.8	167.6
22	Sep-97	145	163.6	171.4
23	Oct-97	169	173.4	181.9
24	Nov-97	174	188.6	190.7
25	Dec-97	212	204.6	208.9
26	Jan-98	243	229.6	232.4
27	Feb-98	225	256.8	256.1
28	Mar-98	294	281.4	281.7
29	Apr-98	310	303.4	
30	May-98	335		
31	Jun-98	353		

Figure 10.7   Sales and Moving Averages for Insulator Sales

the average of the Apr-96 sales and the two previous and two following sales figures. The other values in the column are calculated by taking averages of five successive months in the same way. For the moving averages based on 7 months, we include sales in the previous 3 and following 3 months. The 7-month moving averages are given in column D, headed MA(7), in Figure 10.7. Moving averages are easily calculated in Excel. For instance, the value in cell C4 in Figure 10.7 is obtained using the formula

$$=AVERAGE(B2:B6)$$

which is then copied into the other cells in column C.

Figure 10.8 shows a run chart of the original data and the 5-month moving averages.

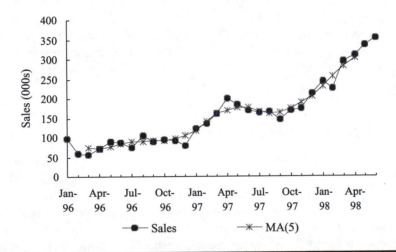

Figure 10.8 Run Chart and Moving Average for Insulator Sales Data

> **Q10.** *What is your forecast of sales in July 1998 based on the moving average trend in Figure 10.7?*

Note that moving averages can be specified in the dialogue window shown in Figure 10.5 by highlighting *Moving average* and specifying the appropriate *Period*; for instance a period of 5 for MA(5) and 7 for MA(7). However, Excel puts the moving average at the end of the period rather than in the centre, as was done in Figure 10.7. It will, therefore, be necessary to edit the *Datasheet* in the chart to produce the plot given in Figure 10.8.

> **Q11.** *For the Pine Products sales data in Figure 10.3, do you think that a linear trend line will be appropriate?*

Q12. *Would it be advisable to use regression to estimate the trend in the Pine Products sales data? Why, or why not?*

To summarise, for a time series that consists of a trend only, a regression model can be used to identify this trend. For some series, and especially when a seasonality component is present, the trend may be difficult to describe by a simple regression function. In this case a moving average can be used. It will smooth out the random variations and reflect the underlying trend in the data. When more periods are averaged, the smoother the moving averages become. However, there will be fewer of them and they will concentrate more and more in the centre of the data.

10.4   Estimating Seasonal Effects

Moving averages can be used to estimate seasonal effects, provided that the number of periods in the moving average is equal to the number of seasons. The size of the seasonal effects can be estimated by comparing the actual figures with the moving averages. This can be done by taking the difference between the actual value and the moving average to give an *additive* measure of the seasonal variation. It is an acceptable way of measuring the seasonal effects if the size of these effects is reasonably constant through the data. However, if there is evidence from the run chart that the size of the seasonal effect is related to the size of the data values, it may be more sensible to take the *ratio* of the data divided by the moving average to give a *multiplicative* factor. Alternatively, in this situation, the size of the seasonal effects will be additive if the logarithms of the data are analysed, rather than the data itself.

The Plutomania attendance data shown in Figure 10.2 are now analysed in Figure 10.9 using 7-day moving averages and additive seasonal factors. The differences between the attendance values and the MA(7) values, given in the final column in Figure 10.9, are the additive seasonal factors. They are set out again in a two-way table of day against week in Figure 10.10. Notice that, due to the centring of the moving averages, there are no seasonal factors for the first 3 and last 3 days.

Q13. *Complete week 4 in Figures 10.9 and 10.10, and verify that the averages in the final column of Figure 10.10 are correct.*

Q14. What do the averages in the final column of Figure 10.10 tell you?

Q15. What should be the mean of the averages in the final column? Check how close it is in this case.

Week	Day	Attendance	MA(7)	Attendance-MA
1	Mon	439		
	Tues	442		
	Wed	514		
	Thur	651	760.1	-109.1
	Fri	717	760.9	-43.9
	Sat	1379	770.1	608.9
	Sun	1179	776.0	403.0
2	Mon	444	772.3	-328.3
	Tues	507	771.4	-264.4
	Wed	555	756.0	-201.0
	Thur	625	758.6	-133.6
	Fri	711	759.7	-48.7
	Sat	1271	759.0	512.0
	Sun	1197	758.7	438.3
3	Mon	452	759.0	-307.0
	Tues	502	758.9	-256.9
	Wed	553	764.6	-211.6
	Thur	627	764.7	-137.7
	Fri	710	751.4	-41.4
	Sat	1311	742.0	569.0
	Sun	1198	744.9	453.1
4	Mon	359	750.1	
	Tues	436	760.0	
	Wed	573	757.4	
	Thur	664	749.4	
	Fri	779		
	Sat	1293		
	Sun	1142		

Figure 10.9   Estimating Seasonal Factors

	Week 1	Week 2	Week 3	Week 4	Average
Mon		-328.3	-307.0		-342
Tues		-264.4	-256.9		-282
Wed		-201.0	-211.6		-199
Thur	-109.1	-133.6	-137.7		-116
Fri	-43.9	-48.7	-41.4		-45
Sat	608.9	512.0	569.0		563
Sun	403.0	438.3	453.1		431

Figure 10.10   Additive Seasonal Factors

We can now use these seasonal factors to forecast daily attendances in future weeks. First we have to project the trend. However, it is clear from the plot in Figure 10.2 that there is no trend in the attendance data, and the moving average in Figure 10.9 is stable at a level of about 750 per day.

Using an average attendance of 750 per day, we can now adjust any individual day's forecast by applying the relevant seasonal factor from Figure 10.10. For example, looking at Mondays, they are typically 342 *below* average, and so *subtracting* 342 from the average level will adjust for this difference. Likewise, adding 563 to the average level on any Saturday will incorporate the "Saturday effect."

> **Q16.** *What will be the forecast for next Tuesday's actual attendance?*

Another use for the seasonal factors is to *deseasonalise* the data. This is where we adjust the actual data values to *take out* the effect of the particular day of the week. This again allows us to see the underlying pattern in the data. This process is sometimes referred to as *seasonal adjustment*. For example, adding 342 to any Monday figure, or subtracting 563 from any Saturday figure, will seasonally adjust for that particular day. Figure 10.11 shows the data both before and after seasonal adjustment.

The seasonally adjusted figures are slightly more variable than the 7-day moving averages. This is because the moving average smoothes out **both** the seasonal and random variations whereas the seasonal adjustment only removes the seasonal component; the random variations are still included, although here they are relatively small.

Seasonality in Pine Products

The run chart in Figure 10.3 indicates clear seasonality in the sales of Pine Products, as can be seen in the actual sales figures in Figure 10.12. Therefore, to forecast future sales, we first need to estimate the size of the quarterly seasonal variations.

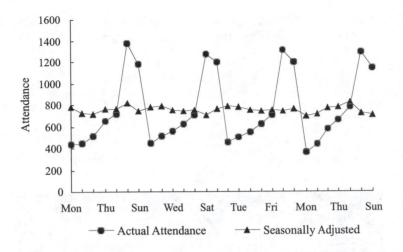

Figure 10.11   Attendance Data after Seasonal Adjustment

Quarter	1995	1996	1997	1998
January – March	36	50	38	54
April – June	4	12	20	34
July – September	28	30	30	48
October - December	52	56	68	84

Figure 10.12   Quarterly Sales ($mil) for Pine Products

Comparing the data values with the moving averages poses a problem when there is an even number of seasons, because the moving averages now fall between the middle two of the figures being averaged. For example, the middle of spring, summer, autumn, and winter is midway between summer and autumn. Before we can work out the seasonal effects, we have to align the moving averages with the data by re-averaging neighbouring MA(4) values using an additional MA(2) average. This is demonstrated in the spreadsheet shown in Figure 10.13.

Notice that the values in cells A2:F17 are positioned at the top or bottom of the cell to show the alignment of the moving averages. The important point is that the first moving average in column D does not align with the data in column C, but the second moving average in column E does. The difference between columns C and E, given in column F, determines the additive seasonal factors.

The additive seasonal factors, copied from column F of Figure 10.13, are given in Figure 10.14.

Q17. *Calculate the overall quarterly factors in Figure 10.14.*

Q18. *Based on the moving averages in Figure 10.13, how would you describe the trend in sales of Pine Products over the last 4 years?*

Q19. *What would be your estimate of the underlying trend in the first quarter of 1999? On what basis did you arrive at this forecast?*

	A	B	C	D	E	F
	Year	Quarter	Sales	MA(4)	MA(2)	Sales - MA(2)
1						
2	1995	Jan-Mar	36			
3		Apr-Jun	4	30.0		
4		Jul-Sep	28	33.5	31.75	-3.75
5		Oct-Dec	52	35.5	34.50	17.50
6	1996	Jan-Mar	50	36.0	35.75	14.25
7		Apr-Jun	12	37.0	36.50	-24.50
8		Jul-Sep	30	34.0	35.50	-5.50
9		Oct-Dec	56	36.0	35.00	21.00
10	1997	Jan-Mar	38	36.0	36.00	2.00
11		Apr-Jun	20	39.0	37.50	-17.50
12		Jul-Sep	30	43.0	41.00	-11.00
13		Oct-Dec	68	46.5	44.75	23.25
14	1998	Jan-Mar	54	51.0	48.75	5.25
15		Apr-Jun	34	55.0	53.00	-19.00
16		Jul-Sep	48			
17		Oct-Dec	84			

Pine Products — Sheet1 / Sheet2 / Sheet3

Figure 10.13   Pine Products Spreadsheet

	1995	1996	1997	1998	Average
Jan-Mar		14.25	2.00	5.25	
Apr-Jun		-24.50	-17.50	-19.00	
Jul-Sep	-3.75	-5.50	-11.00		
Oct-Dec	17.50	21.00	23.25		

Figure 10.14   Additive Seasonal Factors

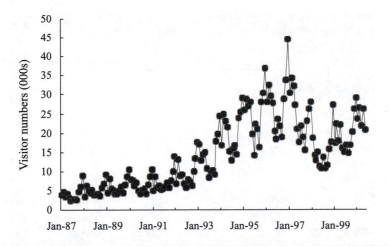

Figure 10.15   Asian Visitors to New Zealand, 1987–2000

Asian Tourism in New Zealand

The monthly number of Asian visitors to New Zealand from January 1987 to May 2000 is given in Figure 10.15. The figures refer primarily to visitors from China, Hong Kong, India, Indonesia, Malaysia, Singapore, South Korea, Taiwan, and Thailand, but exclude Japan.[1]

> *Q20. Accurate forecasts of tourist numbers are important in most countries. Why?*
>
> *Q21. What are the main features of the time series given in Figure 10.15?*

There is clearly a seasonal effect in the tourism numbers in Figure 10.15, with numbers highest in the (southern hemisphere) summer months. However, the effect increases over time, rather than staying constant. Consequently we should either analyse the logarithms of tourist numbers using additive seasonal factors, or calculate multiplicative factors by taking the ratio of the tourist number to the moving average. We will demonstrate the second approach, which is shown in the Excel spreadsheet in Figure 10.16. Only the data for the first 12 months and for 1999 and 2000 are shown.

The underlying trend in the data can be estimated using moving averages, as before. Since there is a 12-month seasonal effect, MA(12) followed by MA(2) has

---

[1]   Source: International Visitor Arrivals to New Zealand, *Statistics New Zealand*, monthly booklet, 1987–2000.

	A	B	C	D	E
1	Month	Visitors	MA(12)	MA(2)	Visitors / MA(2)
2	Jan-87	3,816			
3	Feb-87	4,734			
4	Mar-87	3,492			
5	Apr-87	4,080			
6	May-87	3,042			
7	Jun-87	2,352	4,110		
8	Jul-87	2,616	4,084	4,097	0.639
9	Aug-87	2,840	4,215	4,150	0.684
10	Sep-87	2,608	4,332	4,273	0.610
11	Oct-87	4,752	4,354	4,343	1.094
12	Nov-87	6,140	4,536	4,445	1.381
13	Dec-87	8,844	4,664	4,600	1.923
146	Jan-99	17,200	17,480	17,348	0.991
147	Feb-99	22,282	17,799	17,640	1.263
148	Mar-99	17,797	18,223	18,011	0.988
149	Apr-99	21,931	18,604	18,414	1.191
150	May-99	16,070	19,340	18,972	0.847
151	Jun-99	15,026	19,498	19,419	0.774
152	Jul-99	16758	20,035	19,766	0.848
153	Aug-99	14573	20,384	20,210	0.721
154	Sep-99	16692	20,708	20,546	0.812
155	Oct-99	20244	21,063	20,885	0.969
156	Nov-99	26264	21,438	21,250	1.236
157	Dec-99	29136			
158	Jan-00	23647			
159	Feb-00	26475			
160	Mar-00	21675			
161	Apr-00	26190			
162	May-00	20570			

Figure 10.16   Tourism Spreadsheet

Month	Jan	Feb	Mar	Apr	May	Jun	Jul	Aug	Sep	Oct	Nov	Dec
Average	1.00	1.25	1.04	1.04	0.80	0.68	0.81	0.79	0.68	1.04	1.24	1.58

Figure 10.17   Multiplicative Seasonal Factors

been used in columns C and D in Figure 10.16, and the ratio of the data to the MA(2) values is calculated in column E. From the figures in column E the average monthly multiplicative seasonal factors can be calculated; these are given in Figure 10.17.

> **Q22.** *How do you interpret the seasonal effects in Figure 10.17? What do you conclude from these values?*

10.5   Exponential Smoothing

When a moving average is used to forecast the next period, the main disadvantage is that only the last few figures are used in the forecast (and given equal weight), while all previous figures are ignored. In contrast, *exponential smoothing* uses all previous data, but gives more weight to the most recent data and less and less to data further back in time. The exponentially smoothed forecast made at any period gives a specified weight (a proportion that is usually denoted by α) to the most recent value and the weights *damp down* by a factor $1 - \alpha$ for each period that we go back in time. As the weights are decreasing by a constant factor $1 - \alpha$, they are decreasing exponentially, hence the name of this type of forecasting.

Consider, for example, the insulator sales data in cells B2:B7 in Figure 10.7, which is reproduced in Figure 10.18. In a calculation of the exponentially weighted average based on $\alpha = 0.8$, the most recent value (June) would carry a weight of 0.8 and the weights would decrease progressively by a *damping factor* of 0.2 for each previous month as shown.

Using the weights in Figure 10.18, the forecast for July (F_{JUL}) would be:

$$F_{JUL} = (0.8 \times 87) + (0.16 \times 90) + (0.032 \times 71) + (0.0064 \times 57)$$
$$+ (0.00128 \times 60) + (0.000256 \times 98) = 86.7$$

This is a forecast for July onwards and is updated month by month as each new sales figure is available. For example, if the July sales figure is 75, the updated forecast for August (and all subsequent months) is

$$F_{AUG} = (0.8 \times 75) + (0.16 \times 87) + (0.032 \times 90) + (0.0064 \times 71)$$
$$+ (0.00128 \times 57) + (0.000256 \times 60) = 77.3$$

Month	Jan	Feb	Mar	Apr	May	Jun
Sales	98	60	57	71	90	87
Weight	0.0013×0.2 = 0.000256	0.0064×0.2 = 0.00128	0.032×0.2 = 0.0064	0.16×0.2 = 0.032	0.8×0.2 = 0.16	0.8

Figure 10.18   Exponential Smoothing with $\alpha = 0.8$

This seems to be a rather complicated calculation to have to do repeatedly, but fortunately the exponential nature of the weights means that once a forecast has been calculated (for July, say) that forecast is easily updated to create next month's forecast. The updating procedure is simply

New Forecast = α × Latest Actual Value + $(1 - \alpha)$ × Previous Forecast

i.e.,

$$F_{AUG} = 0.8 \times \text{July sales} + 0.2 \times F_{JUL} = 0.8 \times 75 + 0.2 \times 86.7 = 60 + 17.3 = 77.3$$

How the forecasts vary month by month depends entirely on the chosen value of α. With a very low value of α, all previous observations receive almost the same weight, whereas with a very high value of α, most of the weight attaches to the most recent one or two observations. As a result, a low value of α means that each new sales value does not have much impact, and the previous forecast is only slightly modified, resulting in very stable forecasts that take a long time to adjust to sudden changes in the data. In contrast, a high value of α will give much more erratic forecasts that react more quickly to changes in the data. In the extreme, a value of $\alpha = 1$ will put all the weight on the latest observation, and all values prior to that are ignored.

In practice, a compromise is required between these extremes, with a value of α usually less than 0.5.

✍ *Exponential Smoothing in Excel*

The calculation of an exponentially smoothed forecast is easily done in Excel. There is a data analysis tool that automatically works out a series of exponentially smoothed forecasts for a specified damping factor. To illustrate this we have used the monthly insulator sales data shown in Figure 10.1. As already stated, the data we used above in Figure 10.18 are just the first 6 months of this data, and you can see the forecasts for July and August that we calculated above appearing in cells C8 and C9 of the spreadsheet in Figure 10.19. We have again used a damping factor of 0.2 corresponding to $\alpha = 0.8$.

The data are first entered into the spreadsheet in columns A and B. Only the sales data in column B are strictly necessary; the months in column A are included for ease of reference. The exponentially smoothed forecasts in column C are calculated using *Exponential Smoothing* in the *Data Analysis* section of the *Tools* menu. The dialogue box shown in Figure 10.20 requires the location of the input data (B2:B31), a *Damping factor* (0.2), and the location of the output (the first cell of the column containing the output, C2). The values in column C are then returned after pressing OK. They are shown here rounded to one decimal place.

The errors in column D in Figure 10.19 are given by subtracting column C from column B. The percent error in column E is given by dividing the absolute size of the error (ignoring the sign) in column D by the sales in column B and multiplying by 100 to express it as a percentage, so that the formula in E3 is

= ABS(D3)*100/B3

Finally, the mean absolute percentage error (MAPE) is calculated in cell E32 as the average of column E.

	A	B	C	D	E
1	Month	Sales	Forecast	Error	%|Error|
2	Jan-96	98	#N/A		
3	Feb-96	60	98.0	-38.0	63.3
4	Mar-96	57	67.6	-10.6	18.6
5	Apr-96	71	59.1	11.9	16.7
6	May-96	90	68.6	21.4	23.8
7	Jun-96	87	85.7	1.3	1.5
8	Jul-96	75	86.7	-11.7	15.7
9	Aug-96	104	77.3	26.7	25.6
10	Sep-96	89	98.7	-9.7	10.9
11	Oct-96	95	90.9	4.1	4.3
12	Nov-96	93	94.2	-1.2	1.3
13	Dec-96	80	93.2	-13.2	16.5
14	Jan-97	124	82.6	41.4	33.3
15	Feb-97	135	115.7	19.3	14.3
16	Mar-97	162	131.1	30.9	19.0
17	Apr-97	200	155.8	44.2	22.1
18	May-97	185	191.2	-6.2	3.3
19	Jun-97	170	186.2	-16.2	9.5
20	Jul-97	163	173.2	-10.2	6.3
21	Aug-97	167	165.0	2.0	1.2
22	Sep-97	145	166.6	-21.6	14.9
23	Oct-97	169	149.3	19.7	11.6
24	Nov-97	174	165.1	8.9	5.1
25	Dec-97	212	172.2	39.8	18.8
26	Jan-98	243	204.0	39.0	16.0
27	Feb-98	225	235.2	-10.2	4.5
28	Mar-98	294	227.0	67.0	22.8
29	Apr-98	310	280.6	29.4	9.5
30	May-98	335	304.1	30.9	9.2
31	Jun-98	353	328.8	24.2	6.8
32				MAPE =	14.7

Sheet1 / Sheet2 / Sheet3

Figure 10.19 Exponential Smoothing in Excel

Notice the way that the exponential smoothing starts off. There is no forecast for January 1996 and to get the calculations underway we assume that the forecast for February (C3) is the actual January sales (98). From the actual and forecast values for February (60 and 98) we calculate the March forecast as $0.8 \times 60 + 0.2 \times 98 = 67.6$, and so on. Finally, the forecast for July 1998 would be $0.8 \times 353 + 0.2 \times 328.8 = 348.2$.

The "best" value of α for any set of data is found by trial and error, and is the one that gives the smallest overall error for the data as measured by the mean absolute percentage error (MAPE). For $\alpha = 0.8$, we have MAPE = 14.7%. A straight-forward approach is to use repeatedly the *Exponential Smoothing* procedure in

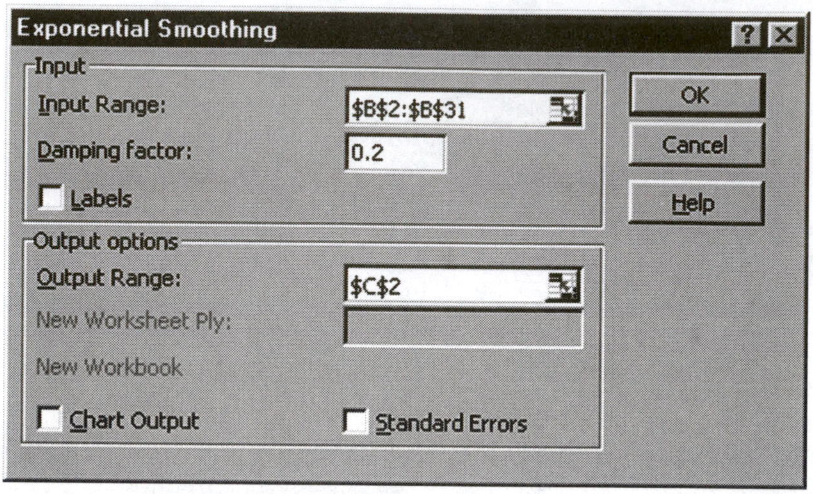

Figure 10.20   Exponential Smoothing Dialogue Box

Data Analysis for different damping factors. Alternatively, a more efficient approach for more experienced users of Excel is to amend the formulae in column C to reference a variable damping factor that is stored in a cell of the spreadsheet. Then, by repeatedly changing the contents of this cell, the effect of alternative damping factors can be seen on the average percentage error.

Figure 10.21 shows how the forecasts compare with the actual sales values.

> ***Q23.*** *You can see that the forecast for each period tends to be lower than the actual sales when the sales are increasing, and higher than actual sales when sales are decreasing. Is there a reason for this or is it just chance?*

The fact that there seems to be an upward trend in the data (particularly after 1996) will have implications for the accuracy of any exponential smoothing forecast. Specifically, the forecasts will lag behind the data because they are averages of previous values when the series was at a lower level. Hence there is a lag or bias in the forecasts. Similarly, with a falling trend, the forecasts would tend to be greater than the actual values. Furthermore, the smaller the value of α the greater is the weight given to "older" data, and so the more effect this will have. Thus exponential smoothing with $\alpha = 0.1$ will have an even greater lag (error) than with $\alpha = 0.8$.

Whenever we use smoothing techniques to forecast data that has a consistently rising or falling trend, we should amend the procedure to allow for the trend and so remove the lag effect. One way of doing this, using a method of differencing, will be described in Section 10.6.

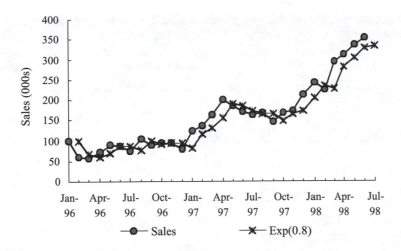

Figure 10.21   Insulator Sales Data and Exponential Smoothing

 Asian Tourism

To illustrate the use of exponential smoothing on seasonal data, we will further examine the data in Figure 10.15 showing the number of Asian visitors to New Zealand since 1987. The process of exponential smoothing can be extended to include an additional smoothing constant to estimate, and continually update, the seasonal factors. However, a simpler approach is to estimate appropriate seasonal factors as described in Section 10.4 and use these to deseasonalise the data before applying exponential smoothing. The exponentially smoothed forecasts are then adjusted (reseasonalised) to incorporate the seasonal effects. This procedure is shown in the Excel spreadsheet in Figure 10.22.

The Excel extract in Figure 10.22 shows the first and last years of the tourism data, with the actual number of visitors in column B. These figures are deseasonalised in column C using the factors calculated in Figure 10.17. The deseasonalised values are exponentially smoothed in column D using $\alpha = 0.5$. The forecasts are then seasonally adjusted (column E) to give a forecast number of visitors in each month. A reasonably high value of α is necessary to keep up with the rises and falls in the data. Subsequent experimentation shows that $\alpha = 0.5$ is the best value for this set of data, leading to the forecast errors in column F and the percentage errors in column G. The mean absolute percentage error (MAPE) is 10%.

> **Q24.** *From Figure 10.22, what is the forecast number of Asian visitors in June 2000?*
>
> **Q25.** *How would you obtain the forecasts for the remaining 6 months of 2000?*

	A	B	C	D	E	F	G
	Month	Visitors	De-season	Exponential	Seasonally	Error	%|Error|
2				($\alpha = 0.5$)	Adjusted		
3	Jan-87	3,816	3816	#N/A			
4	Feb-87	4,734	3787	3816	4770	-36	0.8
5	Mar-87	3,492	3358	3802	3954	-462	13.2
6	Apr-87	4,080	3923	3580	3723	357	8.8
7	May-87	3,042	3803	3751	3001	41	1.3
8	Jun-87	2,352	3459	3777	2568	-216	9.2
9	Jul-87	2,616	3230	3618	2930	-314	12.0
10	Aug-87	2,840	3595	3424	2705	135	4.8
11	Sep-87	2,608	3835	3509	2386	222	8.5
12	Oct-87	4,752	4569	3672	3819	933	19.6
13	Nov-87	6,140	4952	4121	5110	1030	16.8
14	Dec-87	8,844	5597	4536	7167	1677	19.0
159	Jan-00	23647	23647	19708	19708	3939	16.7
160	Feb-00	26475	21180	21678	27097	-622	2.3
161	Mar-00	21675	20841	21429	22286	-611	2.8
162	Apr-00	26190	25183	21135	21980	4210	16.1
163	May-00	20570	25713	23159	18527	2043	9.9
164	Jun-00			24436	16616		
165						MAPE =	10.0

Figure 10.22   Exponential Smoothing on Deseasonalised Data

A run chart of the forecast and actual number of visitors over the last 4 years is given in Figure 10.23. It can be seen that the forecasts follow fairly accurately the variations in the actual numbers, caused both by local trends in the data (year by year variation) and seasonal effects (monthly variation). However, despite the relative accuracy of the forecasts, this example also illustrates some of the difficulties, and potential pitfalls, involved in trying to forecast the future. What would be a forecast of Asian tourism numbers in 5 years' time? What forecast would you have made at the end of 1996 for the tourism numbers in February 2000? Could the Asian financial crisis, and its effect on tourism numbers, have been predicted? The further ahead we attempt to forecast, the greater will be the potential impact of major events such as these and, as a consequence, the less accurate such forecasts will be.

10.6   Differencing

In Figure 10.21 the recent exponentially smoothed forecasts tended to lag behind the data. This is due to the consistent upward trend in the data during the last year. One way of overcoming this is to adjust the data by a process of *differencing*, which will remove the trend.

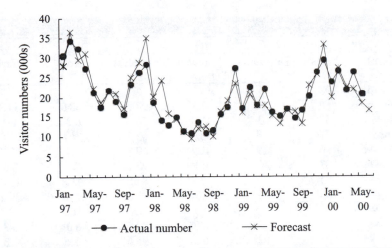

Figure 10.23   Actual and Forecast Numbers of Asian Visitors, 1997–2000

The changes from one value to the next are called *first* differences, and are calculated by subtracting each value from the next one. If necessary, *second* differences can also be calculated by applying the same process to the first differences. First differences will eliminate a linear trend, whereas second (or even higher) differences may be necessary to eliminate nonlinear trends.

Insulator Sales

The plot in Figure 10.24 gives the first differences for the insulator sales data. This series no longer has a systematic trend, although there is a hint of an upward trend in the latter part of the data. Further, the variability in the differenced data appears to be fairly constant over time. Time series with no systematic trend or changes in variability, and where any seasonal variation has been removed, are called *stationary* series. If we assume that the differenced data is stationary, we

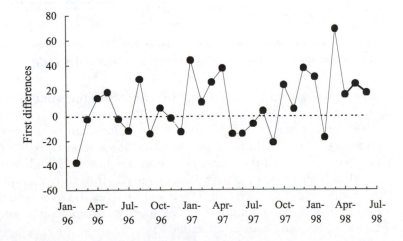

Figure 10.24   First Differences of Insulator Sales Data

	A	B	C	D	E	F	G
	Month	Sales	Difference	Forecast Difference	Forecast Sales	Error	%|Error|
3	Jan-96	98	0	#N/A			
4	Feb-96	60	-38	0.0	98.0	-38.0	63.3
5	Mar-96	57	-3	-3.8	56.2	0.8	1.4
6	Apr-96	71	14	-3.7	53.3	17.7	25.0
7	May-96	90	19	-1.9	69.1	20.9	23.3
8	Jun-96	87	-3	0.1	90.1	-3.1	3.6
9	Jul-96	75	-12	-0.2	86.8	-11.8	15.8
10	Aug-96	104	29	-1.4	73.6	30.4	29.2
11	Sep-96	89	-15	1.7	105.7	-16.7	18.7
12	Oct-96	95	6	0.0	89.0	6.0	6.3
13	Nov-96	93	-2	0.6	95.6	-2.6	2.8
14	Dec-96	80	-13	0.4	93.4	-13.4	16.7
15	Jan-97	124	44	-1.0	79.0	45.0	36.3
16	Feb-97	135	11	3.5	127.5	7.5	5.5
17	Mar-97	162	27	4.3	139.3	22.7	14.0
18	Apr-97	200	38	6.5	168.5	31.5	15.7
19	May-97	185	-15	9.7	209.7	-24.7	13.3
20	Jun-97	170	-15	7.2	192.2	-22.2	13.1
21	Jul-97	163	-7	5.0	175.0	-12.0	7.4
22	Aug-97	167	4	3.8	166.8	0.2	0.1
23	Sep-97	145	-22	3.8	170.8	-25.8	17.8
24	Oct-97	169	24	1.2	146.2	22.8	13.5
25	Nov-97	174	5	3.5	172.5	1.5	0.9
26	Dec-97	212	38	3.7	177.7	34.3	16.2
27	Jan-98	243	31	7.1	219.1	23.9	9.8
28	Feb-98	225	-18	9.5	252.5	-27.5	12.2
29	Mar-98	294	69	6.7	231.7	62.3	21.2
30	Apr-98	310	16	13.0	307.0	3.0	1.0
31	May-98	335	25	13.3	323.3	11.7	3.5
32	Jun-98	353	18	14.4	349.4	3.6	1.0
33	Jul-98			14.8			
34						MAPE =	14.1

Figure 10.25   Exponential Smoothing on Differenced Insulator Sales Data

can apply exponential smoothing to the first differences in the same way that we did in Section 10.5.

The Excel extract in Figure 10.25 shows the sales data with exponential smoothing applied to the first differences calculated in column C, rather than the original data in column B. Furthermore, as the series has no trend, we do not need to use such a high value of α as was necessary before to overcome the lag effect. In this case, a damping factor of 0.9 ($\alpha = 0.1$) gives the best results. Thus, in the exponential smoothing dialogue box we specify an input range of C3:C32, a damping factor of 0.9, and an output range D3.

Exponential smoothing now gives a forecast of the first difference (current value − last value), and so a forecast of the actual sales in any month is obtained by adding last month's sales to the exponentially smoothed figure. This is calculated

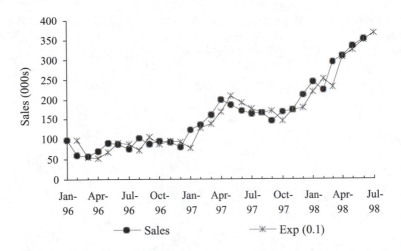

Figure 10.26 Insulator Sales Data and Differenced Exponential Smoothing

in column E. The errors in column F are the differences between the actual sales (column B) and the forecast sales (column E). The %|Error| and MAPE are calculated as before. To find the best value for the damping factor, experimentation with different values of α was undertaken to finally arrive at a value of $\alpha = 0.1$. The resulting MAPE = 14.1% is only slightly better than the MAPE = 14.7% for the undifferenced data. However, we shall see in the next section that the forecasts are better from another aspect.

A graph of the sales data and the exponentially smoothed forecasts is given in Figure 10.26.

> **Q26.** *What would be your forecast for July 98?*
>
> **Q27.** *Compare the plots given in Figures 10.21 and 10.26. What conclusions do you draw?*

Finally, we should stress that differencing is just one way of adjusting for a trend to create a stationary series, and exponential smoothing is just one way of modeling a stationary series. Numerous other models, such as the family of autoregressive models, are available, but beyond the scope of this book.

10.7   Residual Analysis

As in regression, the residuals from the fitted model are a useful indication of how well the model fits the data. We could plot the residuals against the forecasts, but as the data are in time order it will be more useful to plot them against time. Any patterns in the residuals over time will then be clear from this plot and, as before, will indicate an inappropriate model. For example, the plot of the residuals in Figure 10.19 from forecasts based on the raw insulator sales data is shown in Figure 10.27a. The plot of the residuals based on the differenced data is given in Figure 10.27b.

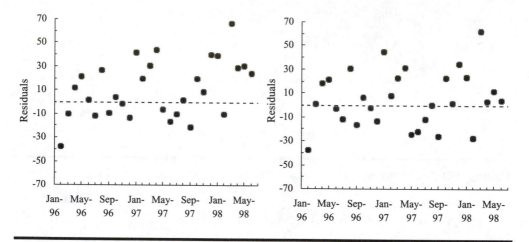

Figure 10.27   Residual Plots for (a) Raw Data, (b) Differenced Data

Q28. What do you conclude from a comparison of Figures 10.27a and 10.27b?

For this particular data set an exponential smoothing model with $\alpha = 0.1$, using first differencing to eliminate the trend, produces a lower MAPE and a more acceptable residual plot. It seems to be the most appropriate model. Note, however, that we have only touched on simple time series models and there are many more models and techniques we could have used that are beyond the scope of this book. It is quite likely that a more sophisticated model would improve the forecasts still further.

10.8   Forecasting Accuracy

One drawback to using MAPE, as described above, for measuring the residuals in a forecasting model is that it tends to overstate the true accuracy of the model.

Q29. *Why do you think this is?*

A more reliable indicator of how good a model is for forecasting is to measure its performance on *new* data, that is, data that have not been used to construct and *calibrate* the model itself. It is a bit like comparing cars that have been tuned to give optimum performance in particular driving conditions. Arguably, it is better to compare them in more general conditions that were not known in advance, rather than in the conditions for which they have been set up.

This implies that we should, if possible, reserve some of our data for testing the model and not use that data in building the model. How much data is used for testing, and how much for model building, will depend largely on how much data is available. When a large amount of data is available, the proportion reserved for testing might be about a third. With less data available, you may need to use proportionately more for model development in order to produce a reliable model.

It is to be expected that the performance of the model on the test data will be worse, often significantly so, than the internal errors on the data used for model development. For example, it is not unusual for the MAPE on test data to be as much as 50% greater than on the development data. After all, if a car has been set up to give good fuel economy under motorway conditions, it will not be surprising that it does not perform nearly so well in urban conditions.

This is essentially the same issue as in Chapter 8 when we examined the margin of error of a regression forecast. We saw there that as we extrapolate further out from the centre of the data, the margin of error increases. Using a regression line to predict beyond the range of the data leads to greater residuals than predicting within the range of the data. Similarly, using a forecasting model to predict the values of new (or "hold out") data that was not used to construct the model is likely to be less accurate than for the data on which the model was built.

10.9 Exercises

1. Which of the following forecasting models will produce the most stable, least variable, forecasts?
   a. exponential smoothing with $\alpha = 0.1$
   b. exponential smoothing with $\alpha = 0.2$
   c. exponential smoothing with $\alpha = 0.9$
   d. exponential smoothing with $\alpha = 1$

2. Using regression analysis to forecast future values of a time series with a seasonal component
   a. will lead to an unbiased estimate of any trend in the data
   b. will underestimate the true trend in the data

    c. will overestimate the true trend in the data

    d. should be avoided as you cannot tell what effect the seasonality will have on the regression line

3. A plot of monthly sales data from January 1997 to June 1999 is given in Figure 10.28. A straight line regression analysis was used to give a forecast sales figure in July 1999 of 200,000 units. Which of the following is **not** true for this method of forecasting?

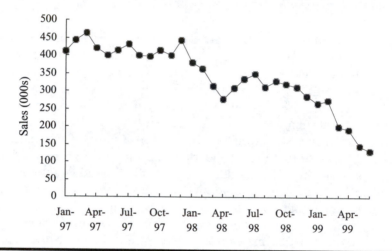

Figure 10.28   Monthly Sales Figures

    a. All past data are given the same weight in deriving the forecast.

    b. The sales figures are not independent of each other.

    c. The forecast is higher than we might reasonably expect.

    d. The sales figures in 1997 have little or no influence on the forecast.

    e. It is not a good forecasting method.

4. Hickson Tools is trying to forecast sales of its P34 chain saw. Monthly sales of chain saws are given in Figure 10.29.

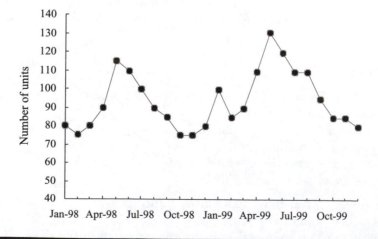

Figure 10.29   Sales of Chain Saw Model P34

a. Do you think there is a trend in the data? Can you see any seasonal patterns?
b. What forecasting methods do you think are appropriate to predict likely sales in January 2000?
c. Explain how you would go about forecasting the sales of the chain saw in January 2000.

5. The sales of videotapes have been recorded over the past 3 years and a plot is given in Figure 10.30. The table shown in Figure 10.31 has been produced using exponential smoothing in Excel.

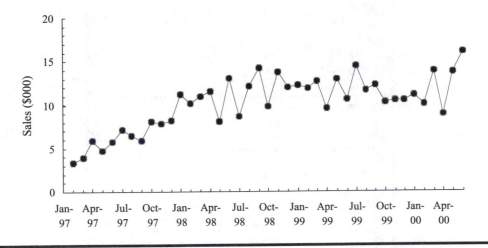

Figure 10.30   Video Sales

a. What are the main components of this time series?
b. What are the forecast sales for July 2000?
c. Comment on how much weight your forecast places on the 1997 data.
d. How would you judge whether this is a good forecasting method?

6. Mildred Homes, the chief accountant, is drawing up a budget for next year's electricity usage. She has available the electricity usage for the past 3 years as shown in Figure 10.32.
a. What are the most important components of this time series?
b. What periodicity should she choose for her moving average?
c. She estimates the total electricity usage next year will be 18,000 kWh. Explain how you would estimate the usage in June 2000 (no calculations are required). Will it be less or more than average monthly usage?
d. Mildred has heard that exponential smoothing is a good forecasting technique, so she is considering it as an alternative, with a smoothing constant of $\alpha = 0.1$. Explain how you would use this approach in this situation.

7. Wintago Hospital has been trying to forecast weekly patient admissions to determine the number of nurses it needs to employ. The admissions over 28 weeks are shown in Figure 10.33.
a. Is there a trend in the data? Can you see any seasonal patterns?

Video sales			
	A	B	C	D
1	Month	Sales	Forecast	Error
2	Jan-97			
3	Feb-97	3.4	#N/A	
4	Mar-97	3.9	3.40	0.50
5	Apr-97	5.9	3.65	2.25
6	May-97	4.7	4.78	-0.08
7	Jun-97	5.8	4.74	1.06
8	Jul-97	7.2	5.27	1.93
9	Aug-97	6.5	6.23	0.27
10	Sep-97	5.9	6.37	-0.47
11	Oct-97	8.1	6.13	1.97
12	Nov-97	7.9	7.12	0.78
13	Dec-97	8.2	7.51	0.69
14	Jan-98	:	:	:
15	:	:	:	:
16	Dec-99	:	:	:
17	Jan-00	11.1	10.69	0.41
18	Feb-00	10.1	10.90	-0.80
19	Mar-00	13.9	10.50	3.40
20	Apr-00	8.9	12.20	-3.30
21	May-00	13.8	10.55	3.25
22	Jun-00	16.1	12.17	3.93

Sheet1 / Sheet2

Figure 10.31   Extract from Excel Spreadsheet for Exponential Smoothing ($\alpha = 0.5$)

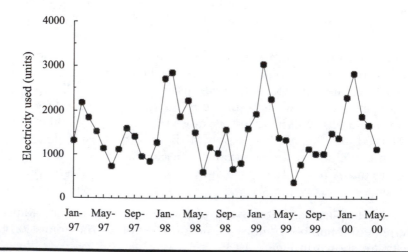

Figure 10.32   Electricity Usage

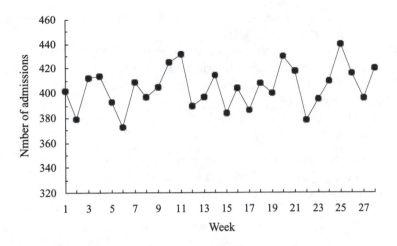

Figure 10.33   Hospital Admissions

	A	B	C	D	E
1	Week	Admissions	MA(3)	Forecast	Error
2	1	402			
3	2	379	397.7		
4	3	412	401.7		
5	4	414	406.3	397.7	16.3
6	5	393	393.3	401.7	-8.7
7	6	373	391.7	406.3	-33.3
8	7	409	393.0	393.3	15.7
9	8	397	403.7	391.7	5.3
10	9	405	409.0	393.0	12.0
11	10	425	420.7	403.7	21.3
12	11	432	415.7	409.0	23.0
13	12	390	406.3	420.7	-30.7
14	13	397	400.7	415.7	-18.7
15	14	415	398.7	406.3	8.7
16	15	384	401.0	400.7	-16.7
17	16	404	391.3	398.7	5.3
18	17	386	399.3	401.0	-15.0
19	18	408	398.0	391.3	16.7
20	19	400	412.7	399.3	0.7
21	20	430	416.0	398.0	32.0
22	21	418	408.7	412.7	5.3
23	22	378	397.0	416.0	-38.0
24	23	395	394.3	408.7	-13.7
25	24	410	415.0	397.0	13.0
26	25	440	422.0	394.3	45.7
27	26	416	417.3	415.0	1.0
28	27	396	410.7	422.0	-26.0
29	28	420		417.3	2.7
30	29			410.7	

Figure 10.34   Hospital Admissions, Forecasts, and Errors

b. What type of forecasting method do you think is appropriate to predict likely admissions in week 29?

c. The hospital admissions data are given in Figure 10.34. A 3-weekly moving average has been calculated, along with the forecasts and errors. Do you think the forecast of 411 admissions for week 29 is realistic? If not, what else would you do?

d. What other forecasting methods do you think might be appropriate for this data set?

8. A sports goods retailer has collected quarterly figures for the value of sales ($000) for all its shops in Victoria. A plot of the data is given in Figure 10.35. They wish to forecast quarterly sales for the year 2000.

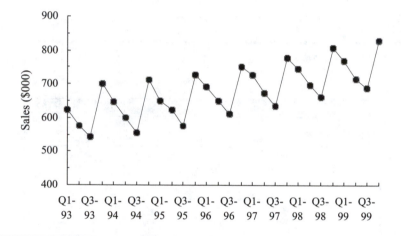

Figure 10.35   Quarterly Sales Data

a. What are the main components of this time series?

b. What method of forecasting would you advise them to use for their quarterly forecasts?

c. What should you do before using exponential smoothing on these data? The quarterly seasonality indices for this set of data are given in Figure 10.36.

Quarter	Seasonality index
1	25.1
2	-25.9
3	-71.3
4	72.1

Figure 10.36   Seasonality Indices

d. Explain what these quarterly seasonality indices mean.

e. If the sales forecast for total sales next year is $3,200,000, estimate the sales for each quarter.

f. How do you think you could determine next year's total sales?

Chapter 11

Statistical Process Control

11.1   Introduction

Statistical process control (SPC) is a collection of management and statistical techniques whose objective is to bring a process into a state of stability or control, and then to maintain this state. All processes are variable and being in control is not a natural state. SPC has been found to be an effective way to improve product and service quality. There are many management issues, as well as statistical techniques, which need to be considered to ensure the successful use of statistical process control. Quite often SPC fails in companies, not because of the technicalities, but because of a poor understanding of the important management issues.

Statistical process control is part of an overall plan to reduce variation and improve business processes. In *The Team Handbook*, Scholtes outlines a blueprint for improvement, which he calls the "Five Stage Improvement Plan." A schematic of the plan is shown in Figure 11.1.

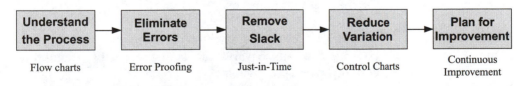

Figure 11.1   The Five Stage Improvement Plan

Statistical process control plays a part in reducing variation and planning for continuous improvement, the last two stages in the improvement plan. Many companies are so involved in fire-fighting problems that they never even get to understand their processes. Other companies that are committed to process improvement (often in the manufacturing sector) are mainly concerned with trying to eliminate errors and removing slack from their systems. Companies that apply

227

SPC and plan for improvement do enjoy substantial benefits from doing so. For example, a packaging company improved the quality of the corrugated board it manufactured by applying statistical process control. First they collected data on the thickness of the board every half hour and plotted a run chart. Then they used SPC to identify some of the process problems, and eliminated them. The thickness of the board became less variable almost immediately. Previously they had been continually adjusting the pressures and temperatures of the corrugating machine without any long-term success, a strategy that in fact only made things worse.

In the next three chapters we shall consider and discuss the following aspects of statistical process control:

- The benefits of reducing variation.
- The effect of tampering with a process. This is often done in practice but usually ends up increasing rather than reducing variation.
- The common cause highway, statistical control, and stability.
- The distinction between special and common causes of variability. Understanding this distinction is the key to reducing variation.
- The construction and use of control charts to help us identify common and special causes.
- The establishment of control charts and their use in monitoring processes.
- Specifications and capability. We need to examine whether a process is capable of meeting customer needs, as determined by process specifications.
- Strategies for reducing variation. Different strategies are needed to reduce common and special causes of variation.

We shall be more concerned with the concepts and management issues involved with SPC, rather than with the technicalities.

11.2   Processes

The aim of statistical process control is to improve processes. What is a process? Processes cover everything that goes on in an organisation. In fact an organisation is made up of hundreds, maybe thousands, of interrelated processes. Anything that goes on is part of some process. Some examples are

Manufacturing	designing, assembling, packaging, maintaining equipment.
Finance	payroll, accounts payable, accounts receivable, data processing, word processing.
Employee relations	hiring new staff, training, salary reviews.
Marketing	advertising, promotion, customer service, customer complaints, selling.
Distribution	order taking, order filling, delivering, inventory management.

A process is made up of three essential elements: inputs, processing system, and outputs. The flow of a process is shown in Figure 11.2.

A process takes inputs and then operates a processing system to use and modify the inputs, adding "value" on the way, to produce useful outputs. The inputs may be materials (steel, electrical components, paint, paper, electrical power, etc.), but could also be people (ideas, work, etc.), equipment (machines, tools, etc.) and

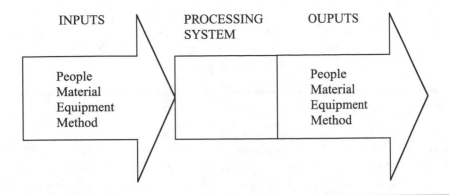

Figure 11.2   Elements of a Process

methods (systems, IT, etc.). Similarly, the outputs produced will fall into one or more of these classes.

Q1. What are the inputs and outputs of the beads experiment?

Q2. How do you think processes are monitored and improved in most organisations? Think of an organisation with which you have been involved (school, university, holiday job, etc.).

Processes produce variable outputs. Variability arises from two sources: the processing system and the variability of materials used in the inputs. Figure 11.3 shows variability in inputs which, when added to the variability from the processing system, produces outputs that are often more variable.

We improve a process by collecting and analysing data from the inputs and outputs, and from studying the process itself. In particular, we concentrate on the

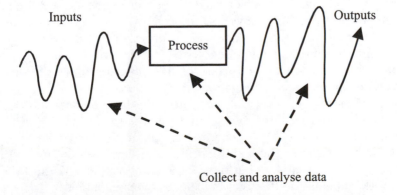

Figure 11.3   Variability in Inputs, Processing System, and Outputs

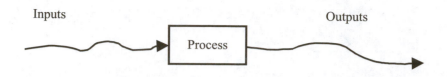

Figure 11.4   An Improved Process

outputs to see if we can identify causes of variability, and then put in place solutions that address them. The problem-solving techniques of Chapter 2 can be used to achieve this aim. By applying such methods, the process will be improved resulting in less variability in the inputs, outputs, and the processing system, as depicted in Figure 11.4.

11.3   Benefits of Reducing Variation

Reducing variation is the key to improving quality, productivity, and profitability. Some of the benefits that can be gained from reducing variability include reduced costs, fewer errors, smaller inventories, shorter set-up times, greater throughput, a more reliable and consistent service, and greater customer satisfaction.

🏭 *Dice Experiment*

To demonstrate the effects of reduced variability on throughput, we will consider a simple manufacturing system where there are five processes: *material supplies, manufacturing, quality control, packing*, and *delivery*. At the end of the system a component is produced, which is then supplied to a customer. A schematic of the system is shown in Figure 11.5.

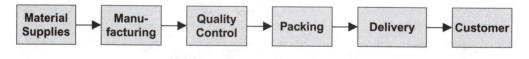

Figure 11.5   The Manufacturing System

The customer requires 35 components every 2 weeks. The production line works 10 days every 2 weeks, so the average daily output needs to be 3.5 components per day. As in most production systems the cost of production not only depends on labour and materials, but also on work in progress (WIP) levels.

We can simulate the manufacturing system by throwing a die at each step of the production process. The face value of the die determines the amount that can be worked on, provided enough material from the previous stage is available. For instance, if a six is thrown, but there are only three items available from the previous stage, then only three items can be worked on.

Three different dice will be used, which differ in the average values and variability of the six sides. These dice are listed in Figure 11.6. The first two dice have the same high variability, but differ in average value. The third die has the same average value as the first die, but much lower variability.

For simplicity, we assume that the supplier of the raw material is operating a *just-in-time* policy such that at the start of each day there are just enough materials,

Die	Face values	Mean	Standard deviation
1	1, 2, 3, 4, 5, 6	3.5	1.87
2	2, 3, 4, 5, 6, 7	4.5	1.87
3	3, 3, 3, 4, 4, 4	3.5	0.55

Figure 11.6 Experimental Dice

taking into account any unused materials from the previous day, to allow for the maximum possible production on that day. So, for example, if there are 3 units of raw materials left over from the previous day, and if the maximum possible production level is 6 per day, then 3 further units of raw materials are made available.

At the beginning of each day, the first throw of the die gives the number of units that are *manufactured* on that day. Any excess of raw materials is held as WIP at the manufacturing stage and is available for use the following day. The second throw of the die determines the number of units that can be handled by *quality control*. That is, it gives the capacity of quality control for that day. If this capacity exceeds the number of units available from manufacturing, then all these units can be passed on to packing, with no WIP. On the other hand, if the capacity is less than the number of units received from manufacturing then only this number can be passed on to packing, with the excess units being WIP at the quality control stage. The die is thrown two more times to represent capacity at the packing and delivery stages, where the same process is repeated so that finally the amount sent to the customer on each day is determined. This whole process is repeated for 50 days and with each of the 3 dice. Our contract is to supply the customer with an average of 3.5 components per day, that is, a total of 175 components over the 50 days.

Q3. What data should we collect from the simulations to compare the different patterns of production?

We suggest that you experience the dice experiment in a group situation. One option is to move material (such as pieces of paper or counters) from one process to another on the throw of a die. Another option is to simulate the experiment using Excel. Below we describe the experiment, using the Excel spreadsheet given in *Dice.xls*.

The spreadsheet in Figure 11.7 shows 5 days of simulated production, using die 1. The results of the 5 days are given in cells A9:V13; the dice above the four processes showing capacity on the last day simulated (in this case day 5). If we step through the results for day 1, given in row 9, we see that there are 6 pieces available to work on in *manufacturing*. However, the 4 thrown by the die for

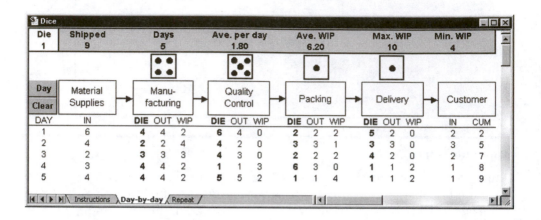

Figure 11.7   Dice Experiment Spreadsheet

manufacturing means that they can only work on 4 of the 6 available pieces. These four are passed on to *quality control*, leaving the other two pieces as WIP. Although a 6 is now thrown in *quality control*, only 4 components are available to be worked on. The four are passed on to *packing* leaving no WIP in *quality control*. Similarly *packing* is only able to work on 2 of the 4 available leaving 2 as WIP, and *delivery* is able to work on 5 but there are only 2 pieces available. At the end of the first day, therefore, the customer receives only two components. Day 2 proceeds in a similar way, but bearing in mind that there is WIP at some stages available from the previous day.

Rows 1 and 2 of the spreadsheet record cumulative progress. Hence, after 5 days of production the customer has been shipped 9 components for an average of 1.80 per day. The total WIP on each of the 5 days is 4, 5, 5, 7, and 10, respectively, giving an average WIP of 6.2 pieces per day, with a daily maximum and minimum WIP of 10 and 4, respectively.

Further days are simulated using the *Day* button in cell A5. Clicking on the button once gives the results for day 6. Repeatedly clicking this button will add further days. In this way the results for 50 days are obtained. The final cumulative figures are given in row 2. We can then repeat the simulation with different dice by clicking on *Clear* in cell A7 and changing the number in cell A2.

We suggest you run the experiment using the Excel spreadsheet *Dice.xls*. Record your results for 50 days of production for each of the *three* dice in Figure 11.8. If you are not experiencing the experiment for yourself, enter the data given in Figure 11.20 into Figure 11.8.

Die	Components shipped	Average WIP per day
1		
2		
3		

Figure 11.8   Results from Simulation Experiment

The *Repeat* worksheet in *Dice.xls* can be used to simulate a number of successive runs of the experiment for each of the three dice. Try running a few experiments to see what sort of variability you get in your results.

> **Q4.** *What conclusions do you draw from Figure 11.8?*

 Kanga Packaging

A manufacturer of cardboard cartons was unable to meet the needs of one of its major customers during a particular period. In order to meet the demand the production manager arranged for the workers to work 2 hours overtime per day.

> **Q5.** *In light of the dice experiment, comment on this decision. What should he now do?*

11.4 Process Adjustment

As the dice experiment has shown (in Figure 11.8), considerable benefits can be achieved by reducing the variability in a process. Many organisations and managers try to achieve these benefits by continually adjusting the process. They will institute changes in the process when the results appear to be on the high side or low side, or set up other procedures for dealing with change. They are attempting to compensate for outputs that *appear* to be drifting away from some target, and by so doing to bring the process back on target. In other words, to exercise what they believe to be appropriate operational or managerial control.

To illustrate some of the effects of process adjustment, we will consider an example based on a real-life manufacturing process.

Eston Tubes

Eston Tubes manufactures aluminum tubing for use in the construction of racing bicycle frames. They produce 10-m lengths of tubing using an extrusion process. As each length is extruded a laser device automatically determines the average diameter of the tubing, which is displayed on a screen. Bernard has been employed to ensure that the extruder produces aluminum tubing of the required 50 mm diameter. He has been told that if the display shows that tubing is off target then he should try to compensate for this by increasing or decreasing the size of the extrusion nozzle, so that the next length of tubing will be acceptable. Hence, if the screen display shows that the diameter is off target then Bernard can move a dial to either increase or decrease the size of the extrusion nozzle. He does this to try to ensure that the next length of tubing is acceptable. Tubing that has a diameter of more than 2 mm from target has to be scrapped.

> **Q6.** *This is an example of an operator using judgment to make regular adjust-ments to a production process. Is this to be encouraged, or can you see any drawbacks to what Bernard is doing?*
>
> **Q7.** *What do you think the long-term effects of such practices might be?*

The consequences of actions such as those used by Bernard can be easily demonstrated by setting up an experiment to investigate what happens to the output from a process when it is subjected to regular adjustment. A mechanical sampling device called a *Quincunx* can be used to demonstrate these effects. A picture of a quincunx is given in Figure 11.9. The device has an adjustable funnel that feeds balls through a pin board into a series of parallel slots. The lines of pins on the pin board are offset so that a ball falling from the funnel will bounce off successive pins with an equal chance of going either left or right each time. Thus there is more chance that the balls will fall in the central slots rather than the outer ones. If we drop a large number of balls, the slots fill up to form the shape of a distribution. Assuming that we keep the funnel fixed in the central position, that is, always targeting the middle slot, the distribution of balls will be close to a normal curve with its mean near to the centre slot. You can see this distribution beginning to form in Figure 11.9.

We can imagine that the middle slot of the quincunx represents the 50-mm target for the extrusion process. Slots to the left of the middle slot represent decreases of 1 mm and slots to the right increases of 1 mm. With the funnel positioned over the centre slot (50 mm), a ball is dropped from the funnel, through the pin board, and into one of the slots. This corresponds to a tube being extruded, with the resulting diameter given by the value of the slot containing the ball. If the funnel stays positioned over the 50-mm slot, this represents the situation where Bernard sets up the extruder to the best of his ability to produce 50-mm tubing, with no subsequent adjustment.

If you have a quincunx with an adjustable funnel then it can be used to imitate Bernard's actions. Alternatively, you can simulate the experiment using the Excel spreadsheet given in *Quincunx.xls*. Using either the quincunx or spreadsheet, drop 25 balls without making any adjustment to the funnel (spreadsheet method 1),

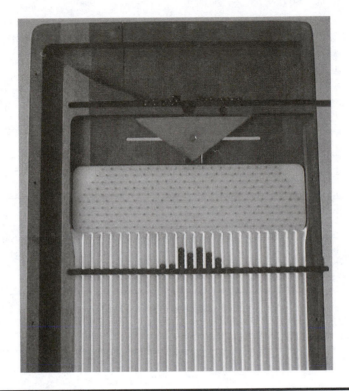

Figure 11.9   Quincunx

and plot your results in Figure 11.10. Alternatively, use the results in Figure 11.21 to answer Q8.

> ***Q8.*** *What proportion of the results will be scrapped, that is, below 48 mm or above 52 mm?*

But this is not what Bernard is doing. He is trying to keep the process on target by continually adjusting the size of the nozzle. Suppose Bernard adjusts the process by moving the nozzle dial down 1 mm for every millimetre above 50 mm, and up 1 mm for every millimetre below 50 mm. For example, if the screen shows 53 mm, he turns the dial down 3 mm for the next tube. Likewise, if the screen displays 48 mm, he turns the dial up 2 mm before the next tube is extruded.

This form of adjustment is called *adjusting for deviation*. It seems a natural way of trying to keep a process on target. After all, if a particular result is below the target, then we should surely make an upward adjustment. We can use the quincunx or spreadsheet to imitate the repeated adjustments made by Bernard by moving the funnel either to the left or right before each ball is dropped, depending on the result of the previous ball. Now drop 25 balls making this form of adjustment (spreadsheet method 2), and plot your results in Figure 11.11. Alternatively, use the results in Figure 11.22 to answer Q9 and Q10.

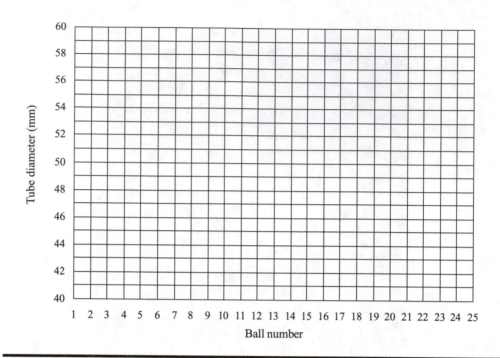

Figure 11.10   No Adjustment

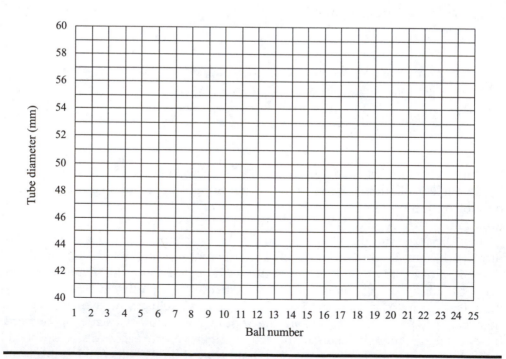

Figure 11.11   Adjusting for Deviation

Q9. *In Figure 11.11, what proportion of tubes will be scrapped?*

Q10. *Has Bernard improved the situation by continually adjusting the nozzle size? Why, or why not?*

Tampering

The type of process adjustment described above is an example of what is known as *tampering*. It can be done by anyone: line worker, manager, and even machines. Such actions are generally misguided, and will usually make the situation worse, not better. If a process has only random variation, tampering with it will introduce even more variability into the process. Overreaction to random variation only increases the variation.

Tampering with a process occurs in many ways. Adjusting for deviation is a common practice, because it is natural to try to reset a process that seems to be deviating from some target. However, the form that the adjustment takes can vary. For example, adjustment might only be made after two or three results have all fallen on one side of the target, or the amount of adjustment might try to compensate for a perceived bias in the output by setting the process "off-target" in the opposite direction to the run of results.

Another common form of tampering is to try to reduce variability by targeting the most recent results. This form of tampering is called *each one like the last*, and is based on the belief that all variation can be controlled, and ultimately eliminated, by trying to produce complete consistency in a process. More realistically, we might only shift to a new level when we get one (or more) values that deviate from the previous target by more than some amount. This form of adjustment is not common in the type of production process that Bernard is controlling.

We can use the quincunx or spreadsheet (method 3) to simulate an *each one like the last* form of tampering. Before each ball is released, the funnel is positioned over the slot into which the previous ball fell.

Q11. *What do you think will happen if we drop a large number of balls using this method of adjustment?*

Q12. *Check whether you are right by using the quincunx or the spreadsheet (method 3) to simulate the effects of "each one like the last." What do you see?*

In most industries tampering is a major source of increased costs. The remedies for dealing with excessive variation are to plot the data, stop tampering, and put in place a scientific approach to reduce variation.

 ## Some Examples of Tampering

Consider the following examples of practices that are reasonably common in many companies. Each can by considered a form of tampering. What form of tampering is involved in each case? What do you think the long-term effects of each action might be?

> **Q13.** *The sales department has a telephone budget of $10,000 per year, but actual expenditure last year was $10,500. The manager decided to reduce next year's budget to $9,500.*

> **Q14.** *The billing department has taken on a new employee who is shown how to process bills by the most recently appointed member of staff (after all it was not so long ago that she was shown the ropes).*

> **Q15.** *In a car spraying process batches of paint are mixed to spray around 30 to 40 cars the same colour. A sample of paint left over from the previous batch is used to match up the colour for the next batch.*

11.5 Common Cause Highway

The results of the red beads experiment varied considerably. This variability:

- Was due entirely to chance—just random fluctuations.
- Had nothing to do with the operators. So it was unfair to penalise them or give them bonuses for the results they achieved as it was outside their control.
- Was a consequence of the *system*, namely there were too many red beads in the raw material. The only way to get better results is to improve the raw material. Improving the system is the job of the manager.

A run chart of the results obtained from running the beads experiment 17 times was given in Figure 1.8, and reproduced in Figure 11.12. Remember, in each

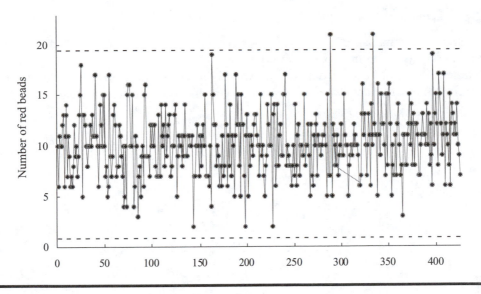

Figure 11.12   Results from 17 Beads Experiments

experiment, 5 operators produce a total of 50 beads (the number in the paddle) for each of 5 days. There are, therefore, 425 individual results in Figure 11.12.

We can see that nearly all the data lie within the dotted lines we have drawn on the run chart. If we repeated the experiment again we would almost certainly get our data falling within these dotted lines. The beads experiment only generates what is known as common cause variation. We shall call the area within these lines the **common cause highway**. We would be surprised if we had a result falling outside the common cause highway. Getting 0, 30, 40, or 50 red beads in the paddle would certainly surprise us. We would suspect that something unusual or *special* had occurred. Even 20 red beads would give us cause to suspect that some special event took place. Note that we have had a value falling outside the common cause highway just twice in our 425 results, which is less than 0.5% of the time.

Almost certainly, therefore, any data value that lies outside the common cause highway is the result of some *special cause of variation* being present. Such values are always worth investigating further.

The key to understanding, and then reducing, the variation in any process is to be able to distinguish between:

- data that fall within the common cause highway, and
- data that fall outside this highway and are almost certainly special causes worthy of separate investigation.

In Sections 1.2, 1.3, and 1.4 we looked at examples where variability in the data caused managers to take inappropriate decisions. Their difficulties were caused by not understanding the difference between common cause variation (data within the highway) and special cause variation. If the variation observed is common cause variation, as it probably was in every case, then actions that treat certain results as in some way special are inappropriate. What is required is a more systemic remedy aimed at reducing the underlying level of variation in the results.

⛭ *Epic Videos*

Heather is the manager of the Chartwell branch of Epic Videos Ltd. All branch managers have recently been urged by head office to improve sales, and so Heather decides to collect data on the weekly sales of videos over several weeks. A run chart of the data she collected is given in Figure 11.13.

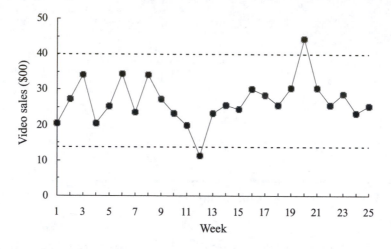

Figure 11.13   Video Sales

Most of Heather's data fall within the common cause highway. Two data points fall outside the highway. There are 2 weeks where specific events caused sales to vary significantly. The low sales in week 12 were probably due to the occurrence of a public holiday that week, and the high sales in week 20 coincided with a TV sales drive. Note, however, that there is considerable variation within the highway. A number of common causes of variation contributed to this variation, such as weather, local sporting events, and TV programmes.

There appear to be two special causes.

- One of them is "good" and when we investigate it we find it is the result of a TV sales drive.
- The other is "bad," but understandable on looking back through the records, as the public holiday probably caused it.

> **Q16.** *What about points inside the highway? Is there any evidence here of any special causes being present? If so, what might indicate this?*

It is important to distinguish between common and special causes of variation. Techniques that enable us to do this comprise the subject area known as *Statistical*

Special causes of variation	Common causes of variation
• Localised in nature, such as a particular supply of raw material, a particular machine, or a specific operator. • Not part of the overall system. • Not always present in the process, as they come from outside the usual process. • To be considered as abnormalities, or as unusual results, or as nonrandom patterns. • Things that can usually be put right by people working on the process, but sometimes needing management intervention. • Causes that typically contribute greatly to variation.	• Those that exist because of the processing system or the way the system is managed. • Present in the process all the time, although their impact varies. • Common to all machines, all operators, and to all parts of the process. • Random fluctuations in process outputs. • Events that individually have a small effect on variation but collectively can add up to quite a lot of variation.

Figure 11.14   Special and Common Causes

Process Control. In Chapter 12 we shall use the data to calculate the boundaries of the common cause highway, to give us charts known as *control charts*. In Chapter 13 we shall develop strategies for reducing variation when we have special or common causes of variation present. Figure 11.14 gives an indication of the type of causes often associated with these two types of variation.

> **Q17.** *The results students obtain in different courses will naturally vary. Think of three or four reasons for this variation, and decide whether you think they are special causes or common causes.*

11.6   Three-Sigma Limits

The arithmetic mean is a measure of location of a set of data and the standard deviation a measure of spread. The arithmetic mean gives the centre line of the common cause highway. If we now add three times the standard deviation to the mean we get the upper boundary of the highway; known as the upper control limit (UCL). Subtracting three standard deviations from the mean gives the lower boundary of the highway, or the lower control limit (LCL). That is, the common cause highway lies within three standard deviations of the mean. We refer to these as three-sigma limits, where sigma refers to the standard deviation. In Chapter 12 we shall be concerned with how we calculate these control limits. For now we shall discuss some of their features.

The first question to ask is why three-sigma? Why have we multiplied the standard deviation by three?

Three-sigma limits have been set in such a way that if a result falls outside these limits then it is almost certainly a special cause. That is, if a result falls outside the common cause highway then we have objective evidence to investigate what has gone wrong. We would suspect that a special cause is present, and we can then look into this. Even with a process with no data outside the control limits, it is still possible that certain results are influenced by special causes; it is just that the results are not sufficiently different for this to be clear. Likewise, there is a small chance of data falling outside the control limits even if no special causes are present. For instance, with the beads experiment it is possible to get 0 or 25 or even 50 red beads in the paddle. But the chances are very small. In fact, with a common cause process we would expect about 2 results out of 1000 to fall outside the *three-sigma* limits purely by chance.

But why *three-sigma* limits? Why not *two-sigma* limits, or *four-sigma* limits?

We shall see in Chapter 13 that we adopt very different strategies for dealing with common and special causes of variation. Common cause variation indicates a system problem. Special cause variation is more indicative of a one-off event that requires immediate and special attention. It is important that we do not confuse these two types of causes, that is, treating a result as if it was a special cause when it is not, and vice versa. We can identify two mistakes that such confusion causes, namely:

- *Interfering too often in the process*, thinking that the problem is a special cause when in fact it belongs to the system. If we use *two-sigma* limits we shall find that we have a lot of results outside the limits (about 1 in 20) when we have a common cause process.
- *Missing important events*, saying that a result belongs to the system when in fact it is a special cause. Not doing anything to find a special cause is another frequent mistake. With *four-sigma* limits we shall miss the opportunity to attack serious special causes.

The use of *three-sigma* limits is thus a compromise, which has proved itself time and time again in practice over the past 50 years. In other words, *it works!*

11.7   Analysis of Patterns

So far we have said that a special cause is signaled by a single result falling off the common cause highway. For a process without any special causes, the results should vary randomly about the centre of the common cause highway. However, specific patterns on the chart may indicate a lack of randomness or lack of stability, and suggest the need to look for special causes. Lloyd Nelson (*Journal of Quality Technology*, 1984) has given a number of other tests based on patterns of points that would be unexpected with a common cause process. Each signals that something unusual is happening. We consider some of these patterns in Figures 11.15 to 11.19. The dotted lines are at *one-sigma*, *two-sigma*, and *three-sigma* above and below the centre line of the common cause highway.

- The situation in Figure 11.15 shows a long run of points below the mean, indicating that the mean has shifted down. A sequence of nine consecutive

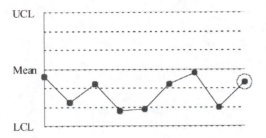

Figure 11.15

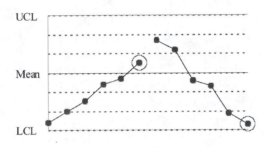

Figure 11.16

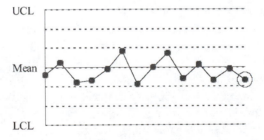

Figure 11.17

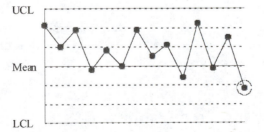

Figure 11.18

points below the mean is a special cause signal. Obviously, the same situation applies when all the points are above the mean.

■ The situation in Figure 11.16, with runs upwards and downwards, indicates a trend in the results. A sequence of six consecutive points steadily increasing or decreasing provides the signal. Such trends may be explained, for instance,

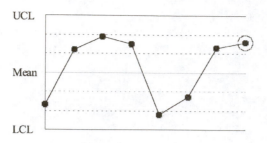

Figure 11.19

by a learning effect being present, a tool wearing, a move to improved materials, or improved operator training.

- In the situation in Figure 11.17 the results are more tightly clustered about the centre line than would be expected with a common cause process. This could be due to incorrectly drawn control limits; due perhaps to not recalculating the limits after some change in procedure had reduced the variability in the process. However, it may also be a result of inspection where results outside the specification limits are not plotted, thus giving a misleading picture of what the process is capable of producing. A sequence of 14 consecutive points within the 1-sigma limits is needed for a signal.

- The systematic saw-tooth effect in Figure 11.18 may be due to a failure to stratify, for instance where the process has been sampled from two different processes alternately. Alternatively, it could be due to overadjustment of the process. A sequence of 14 consecutive points alternating up or down provides a signal.

- Finally, there are no points close to the mean in the situation in Figure 11.19. The cause here is probably failure to stratify. The results above the mean, for instance, may come from one batch of raw material, while the results below the mean from a different batch. There are two separate processes involved. The signal here is a sequence of 8 consecutive points all falling outside the 1-sigma limits.

Note that people are very good at finding patterns in data when none really exist. It is important not to get carried away. The tests above give a set of *objective* criteria to signal the presence of what is likely to be a real underlying pattern or special cause.

11.8   Stability and Predictability

A process is said to be stable, or in *statistical control*, if there are no special causes present. That is, there are no out of control points (all the data are on the common cause highway) and the data are randomly distributed about the centre line (with no detectable patterns).

There are several advantages in having a stable process.

- The process is predictable in the long run, so that the same pattern of variability will occur time and time again and the customer knows what will be received. In contrast, with an unstable process, special causes dominate so it becomes impossible to predict what will happen in the future.

■ Nothing is gained by adjusting a stable process on the basis of its performance. It is relatively easy to make adjustments continuously as information is collected, using, for instance, an automatic feedback system. It is tempting to tamper with the process, as described in Section 11.4. However, adjusting a stable process will only make things worse. Once stable, the process cannot be improved without fundamental systemic changes.

11.9 The 85/15 Rule

It has been established time and time again in companies throughout the world that at least 85% of all problems result from common causes of variation. That is, at least 85% of problems belong to the system, and are thus the responsibility of management. Less than 15% are special causes.

Many believe that the figure is considerably higher than 85%. Dr. W. Edwards Deming, who was a pioneer of the use of statistical process control and widely recognised for his prominent role in the recovery of Japanese industry after the Second World War, put the figure at 96%. The managing director of one prominent New Zealand manufacturing company believes the figure is 100%! That is, all the problems faced by the company in terms of excessive variation in their processes have ultimately been identified as being due to common causes of variation.

This is very different from the way most managers react. They are only too ready to blame people for problems. It is wrong. Most of the time (at least 85%) the problem is exactly like "too many red beads in the raw material"—a system problem.

> *Q18. Is it fair to appraise the performance of the operators involved with producing white beads? What do you think?*

11.10 Implementing SPC

Adding the centre line and the *three-sigma* limits to a run chart gives what is known as a *control chart*. There are two stages in the implementation of an SPC programme, namely the *establishment* of the control chart and the use of the chart to *monitor* the process.

A project team should be responsible for *establishing* the chart. A team approach is useful especially in the identification of special causes. Once the chart is in place, responsibility can then be turned over to the operators. To ensure their cooperation, it is important to have operators involved in the project team.

Basic training is also necessary so that everyone understands the purpose and use of control charts.

Placing a chart at the end of a complex process is ineffective because it is usually only possible to find special causes at the point where they occur. Charts should be placed at critical points within a complex process to make the task easier; that is, the process is best broken down into simple components.

It is much more informative to collect interval data than attribute data. Much less data is required to spot out-of-control conditions if data are measured on some appropriate interval scale rather than being simply recorded on a pass/fail or good/bad basis.

Data collected from a process can have variation that arises from two sources, namely the process itself and the measurement system. If the variation that arises from measurement is so large that it hides the variation in the process, then it will be pointless trying to establish a control chart until the measurement system is reliable. Thus, a preliminary study of the measurement system is advisable.

Once the control limits have been established, further data on the process can be collected and added to the chart. These data can be used for monitoring the process.

Quite often control charts fail in companies, not because of the technicalities, but because of a poor understanding of these and other management issues. The establishment of the control chart is the topic of Chapter 12, and is primarily concerned with how to calculate the control limits given the process data. Monitoring the process, assessing whether it is capable of meeting customer needs, and strategies for improvement are covered in Chapter 13.

11.11 Data for the Dice Experiment

A set of sample results from 50 days of production for each of the three dice is given in Figure 11.20.

Die	Components shipped	Average WIP per day
1	145	15.76
2	197	20.64
3	164	4.94

Figure 11.20   Results from Dice Simulation Experiment

11.12 Data for the Quincunx Experiment

Figures 11.21 and 11.22 give results from dropping 200 balls in a quincunx. Those in Figure 11.21 were obtained when the funnel remained above the centre slot, representing the case of *no adjustment*. The results in Figure 11.22 are for the case when the form of adjustment was *adjusting for deviation*.

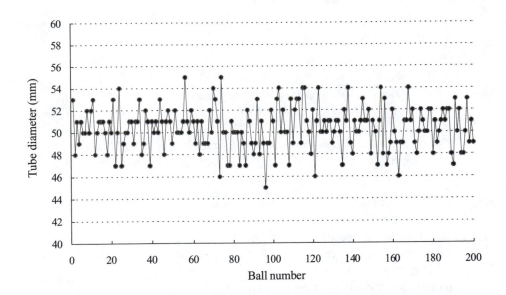

Figure 11.21   No Adjustment

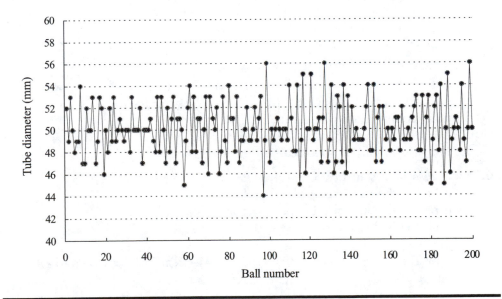

Figure 11.22   Adjusting for Deviation

11.13   Exercises

1. Gertrude decided to give her auditing staff more time to do the accounts this month, since they were late last month. This is an example of
   a. good management practice
   b. a way to reduce the time it takes in the long term
   c. tampering which will increase variation in the long term
   d. tampering which will reduce variation in the long term

e. adjustment that will have no effect on the time to complete the process in the long term

2. Kim is reviewing the last month's sales figures and targets. The sales for the Video 23T were $120,789, well below the target of $150,000. In an attempt to increase sales the following month, she decides to increase the target. This is
   a. an appropriate management reaction to this deviation from budget
   b. an illustration of tampering
   c. a wrong action to take because the target should be adjusted downwards
   d. a wrong action because the target should be adjusted upwards
   e. none of the above

3. Bill looked at the budget for traveling expenses and found that his department had spent only $5,800 of the $7,000 budgeted. He decided to decrease next month's budget to $6,000. This is an example of
   a. good management practice to reduce deviation from budget
   b. a correct use of adjustment
   c. a manager not understanding variability
   d. a lack of initiative among staff

4. The best way to improve the output of a stable process is by
   a. continually adjusting the machine to meet target
   b. making a fundamental change to the process
   c. replacing the operator
   d. installing an automatic process controller

5. A control chart is signaling special causes when there are
   a. two points within one standard deviation of the mean
   b. nine points in a row descending
   c. five points on one side of the centre line
   d. points outside the specification limits

6. A process is in statistical control when
   a. it has a line supervisor
   b. it produces parts within specification limits
   c. both (a) and (b) above
   d. the operator is well trained
   e. it produces parts which are on the common cause highway, and with no patterns
   f. none of the above

7. Which of the following is **not true** (as demonstrated by the dice experiment)?
   a. reducing variation increases throughput
   b. reducing variation decreases inventory
   c. increasing the average production rate increases throughput
   d. increasing the average production rate reduces inventory

8. A control chart that has all its points on the common cause highway is in statistical control
   a. true
   b. false

9. "To control costs, project managers shall report on the costs of all projects that are 10% or more from the budget."
   a. What are your reactions to this policy?
   b. What do you think would be some of the effects of such a policy?

10. Last month it took two extra days to finish the project, so Mary arranged to provide these extra days for this month's project. Mary thought that if they needed two extra days then they might as well have them. Do you believe her decision was sensible? Why, or why not?

11. Applying and enrolling for a degree at a university is a process subject to variability. Some of the variation is due to special causes and some to common causes. Think back to your experience of this process, or some other similar experience. What things were good or bad, and on reflection do you think they were special causes, or common causes?

12. In a certain sense, tampering with a process by adjusting the target depending on recent results is analogous to forecasting a stationary time series. The forecast made at a point in time can be thought of as the target for the next period, that is, the position of the funnel in the quincunx, and how we adjust our forecast in light of the latest result(s) reflects a particular form of tampering.
    a. Think about how we used exponential smoothing in Chapter 10 to give a forecast for a stationary series. What would we do to give an "each one like the last" forecast, and how would we obtain a "no adjustment" forecast?
    b. What happens to the variability of the forecasts as we move from "no adjustment" to "each one like the last?"

Chapter 12

Control Charts

12.1 Introduction

In Section 11.6 we stated that the boundaries of the common cause highway are given by the upper and lower control limits. The *width* of the highway will depend on the amount of variability in the data. The centre line of the highway is given by the mean of the data, and the limits are at three standard deviations each side of this mean. A *control chart* is obtained when these control limits and centre line are added to a run chart of the data. This chapter will be concerned with setting up control charts.

 Nuva Plastics

In order to discuss some of the issues involved, consider the following example from the plastics industry. Ellen Vagner, a line manager with Nuva Plastics, is responsible for the process that produces moulded laundry baskets. Each morning they start the process up at 8:00 a.m., and the die that moulds the basket is heated to 400°C for half an hour, before the first basket is moulded at 8:30 a.m. Production continues throughout the day until the process is closed down at 6:00 p.m. Recently Nuva has received some complaints about the strength of the baskets. It is well known in the industry that strength is mainly related to the density of plastic used in the die; so, they decided to monitor the weight (in grams) of the baskets after they have been moulded. They collected 5 samples spread evenly over a day, and continued collecting data for a period of 25 days. The data are plotted in Figure 12.1. If it appears that there are less than 5 points on any day (for instance on day 3), this is because 2 (or more) values coincide at the same point on the graph.

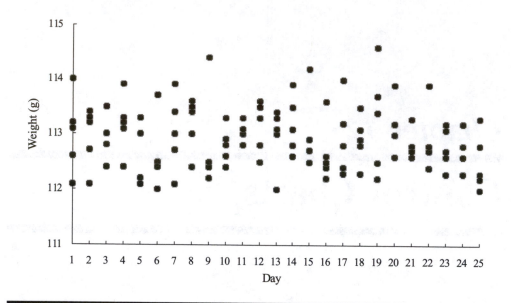

Figure 12.1   Weight of Laundry Baskets

Q1. What can you say about these data? Look in particular at the daily averages and variability.

Q2. How would you calculate an overall measure of variability for these data?

If we look at the daily averages of the data in Figure 12.1 we can see, without doing any calculations, that there is some variation. For example, the average for day 4 is higher than days 2 and 3, and the average for day 25 is lower than the preceding 4 days. Are these daily averages within the common cause highway? That is, is the variation in the daily averages due simply to chance (common cause variation) or are there days when the average is greater or smaller than what might be expected (special cause variation)? Are there any patterns in the way the daily averages vary? A control chart of the daily averages will help us answer such questions.

What can we say about the variation *within* a day? We can see, for instance, that there is a greater range of values in day 19 than in day 11, and that there is very little variation in the 5 samples taken on day 11, but much more variation on day 19. We can construct a control chart of the daily variation, as measured by the daily standard deviation or the daily range. This would give us useful information about whether the weights within our samples are consistent from day to day.

With process data, a small sample of items from the process can usually be taken and measured quite quickly, with the sampling repeated at regular time intervals. With such data we can examine both the variation between and within samples using control charts of the average (X-bar charts) and the range (R-charts). How these charts are obtained, and some of the issues surrounding them, will be discussed in Sections 12.2 and 12.3.

There are situations, however, when it is not possible to collect a number of samples at any one time. Often all we can get is one measurement each day, or week or month. Production figures, sales, and financial data are some examples. The debt recovery and budget deviations data in Sections 1.3 and 1.5 are of this form. The appropriate control chart when we only have individual measurements will be considered in Section 12.4.

Finally, different charts will be required when we collect attribute data. The construction of control charts for proportions and for counts is given in Section 12.5.

12.2   Control Charts for Process Variables

With process variables it is often possible to collect a small sample of data over a relatively short period of time, say a few minutes, and to repeat the sampling at regular intervals, such as every hour. For instance, the shrinkage of a plastic product from an extrusion process, or the weight of blocks of cheese in a dairy factory, or the thickness of a plywood sheet could all be sampled and measured in this way. This sample is called a *subgroup* of data. The variability in the data *within* each subgroup will provide the basis for an accurate estimate of the variability in the process. The control chart of the subgroup means is called the X-bar chart, and the chart of the subgroup ranges is called the R-chart. In previous chapters, we have mainly used the standard deviation to measure variability. However, where small samples of data are being taken regularly, it is much simpler, and just as effective, to use the range as a measure of variability. An R-chart gives a similar picture to one based on the standard deviation, and is much easier to use.

The steps involved in the establishment of the X-bar and R charts are

1. Collect 20 or more subgroups of data. Each subgroup should have the same number of data values (usually 3 to 6, but most commonly 5). Any unusual events should be recorded.
2. For each subgroup, calculate the mean and range.
3. On separate run charts, plot the ranges (R chart) and means (X-bar chart).
4. Calculate the mean of the subgroup ranges, denoted by $\bar{R}$. The *three-sigma* limits for the R chart are given by

$$LCL = D_3\bar{R} \qquad UCL = D_4\bar{R}$$

   where the constants D_3 and D_4 are given in Figure 12.2, and reproduced in Appendix B.5.
5. Look for points outside the control limits on the R chart. If they are identifiable as special causes, they can be removed from the data. The control limits should then be recalculated.

Subgroup size	A₂	D₃	D₄
2	1.880	0	3.267
3	1.023	0	2.574
4	0.729	0	2.282
5	0.577	0	2.114
6	0.483	0	2.004
7	0.419	0.076	1.924
8	0.373	0.136	1.864
9	0.337	0.184	1.816
10	0.308	0.223	1.777

Figure 12.2   Constants for X-bar and R Charts

6. Calculate the mean of the subgroup means. Denote this by $\bar{\bar{x}}$. The *three-sigma* limits for the X-bar chart are given by

$$\text{LCL} = \bar{\bar{x}} - A_2\bar{R} \qquad \text{UCL} = \bar{\bar{x}} + A_2\bar{R}$$

where A_2 is also given in Figure 12.2 and Appendix B.5.

7. Look for points outside the control limits on the X-bar chart. Try to find special causes for these points. If it is possible then these points should be removed and the control limits recalculated.

Note that the constants in Figure 12.2 depend on the size of the subgroup chosen. For example, if we are using subgroups of size 5 then

$$A_2 = 0.577 \qquad D_3 = 0 \qquad D_4 = 2.114$$

These limits are *three-sigma* limits, although this is not apparent from the formulae. Statistical theory tells us that for subgroup sizes less than 10 the range provides a good estimate of the variation in the data. It is not equal to the standard deviation, but a function of it. This function depends on the subgroup size, and this is reflected in the constants given in Figure 12.2.

Nuva Plastics

The plot of the data collected by Ellen Vagner on the weights of moulded laundry baskets is shown in Figure 12.1. The full set of data for 25 days is in cells B2:F26 of the Excel spreadsheet in Figure 12.3. The subgroup means and ranges are calculated in columns G and H. The formula in cell G2 is entered as

$$= \text{AVERAGE(B2:F2)}$$

and the range in cell H2 is calculated from the difference between the maximum and minimum, namely

$$= \text{MAX(B2:F2)} - \text{MIN(B2:F2)}$$

	A	B	C	D	E	F	G	H
	Subgroup	Item1	Item2	Item3	Item4	Item5	Mean	Range
1	1	114.0	112.6	113.2	113.1	112.1	113.0	1.9
2	2	113.2	113.3	112.7	113.4	112.1	112.9	1.3
3	3	113.5	112.8	113.0	112.8	112.4	112.9	1.1
4	4	113.9	112.4	113.3	113.1	113.2	113.2	1.5
5	5	113.0	113.0	112.1	112.2	113.3	112.7	1.2
6	6	113.7	112.0	112.5	112.4	112.4	112.6	1.7
7	7	113.9	112.1	112.7	113.4	113.0	113.0	1.8
8	8	113.4	113.6	113.0	112.4	113.5	113.2	1.2
9	9	114.4	112.4	112.2	112.4	112.5	112.8	2.2
10	10	113.3	112.4	112.6	112.9	112.8	112.8	0.9
11	11	113.3	112.8	113.0	113.0	113.1	113.0	0.5
12	12	113.6	112.5	113.3	113.5	112.8	113.1	1.1
13	13	113.4	113.3	112.0	113.0	113.1	113.0	1.4
14	14	113.9	113.1	113.5	112.6	112.8	113.2	1.3
15	15	114.2	112.7	112.9	112.9	112.5	113.0	1.7
16	16	113.6	112.6	112.4	112.5	112.2	112.7	1.4
17	17	114.0	113.2	112.4	112.3	112.8	112.9	1.7
18	18	113.1	112.9	113.5	112.3	112.8	112.9	1.2
19	19	114.6	113.7	113.4	112.2	112.5	113.3	2.4
20	20	113.9	113.0	113.0	113.2	112.6	113.1	1.3
21	21	113.3	112.7	112.6	112.8	112.7	112.8	0.7
22	22	113.9	112.4	112.7	112.4	112.8	112.8	1.5
23	23	113.2	112.3	112.6	113.1	112.7	112.8	0.9
24	24	113.2	112.8	112.8	112.3	112.6	112.7	0.9
25	25	113.3	112.8	112.0	112.3	112.2	112.5	1.3
26							112.92	1.36

Sheet1 / Sheet2 / Sheet3

Figure 12.3   Nuva Plastics Data

These formulae are then copied to cells G3:H26. The overall mean and the mean range are calculated in cells G27 and H27 using the *AVERAGE* of G2:G26 and H2:H26, respectively.

Q3. What size are the subgroups?

Q4. The subgroup data have been collected over a day. Why do you think this was done?

Q5. What could be a disadvantage of doing this? What would be an alternative?

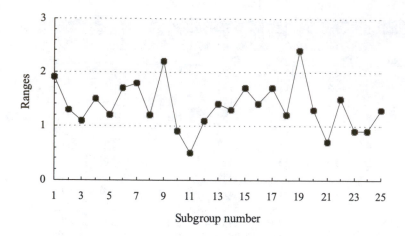

Figure 12.4   R Chart

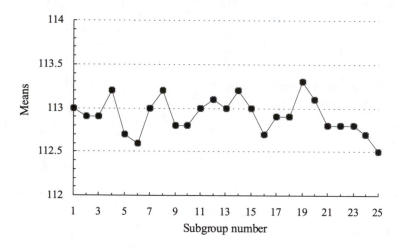

Figure 12.5   X-bar Chart

Q6. *Calculate the control limits for the R and X-bar charts, and plot them on the charts in Figures 12.4 and 12.5. Add the centre lines as well.*

Q7. *Which control chart should you examine first, and why?*

Q8. *What conclusions do you draw from your control charts?*

12.3 Interpreting the X-bar and R Charts

If the X-bar and R charts are both out of control, first examine the R chart. This is because the X-bar chart can be affected by instabilities in the R chart.

The **R chart** monitors the spread or variability of a process and is used to detect changes in the variability of the process. If the variation of the process changes, we would expect the chart to signal a special cause. Although the charts can signal increases or decreases in variability, it is usually points above the UCL that are of most concern. A point above the UCL on the R chart may be due to such things as:

- tool wear
- poorly trained or fatigued operators
- start-up effects
- increased rate of production
- damaged tooling or fixture
- poorer quality raw material or an end of lot
- mixtures of raw material
- automatic controller failure (overadjustment)

For example, suppose a process is sampled initially from the dotted distribution in Figure 12.6. After the 25th data point the process changes to one with greater variability. We are now sampling from the solid distribution. The change is picked up in the R chart after a further five points.

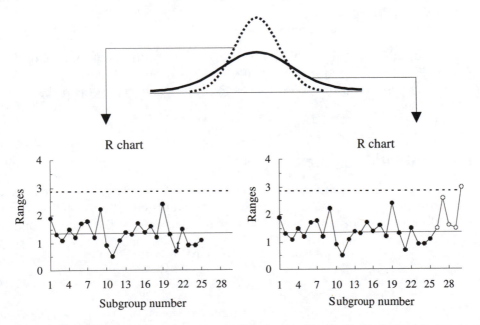

Figure 12.6 Detecting Changes in Variability

The **X-bar** chart (and the chart for individual measurements that will be described in the next section) detects shifts in the general level of the process. A signal from the chart may be due to such things as:

- an adjustment of tooling or fixtures
- a change in raw material, operator, or procedure

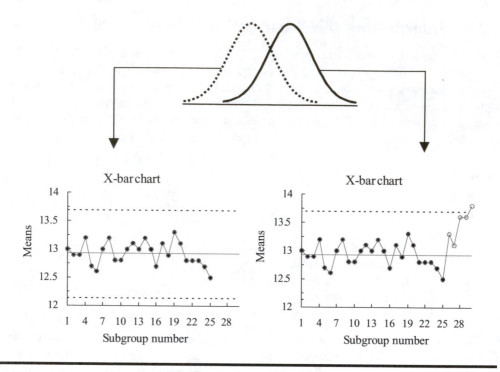

Figure 12.7   Detecting Shifts in the Mean

- a change in operating conditions
- a failure of the measurement system

Again suppose the process is sampled initially from the dotted distribution in Figure 12.7. After the 25th data point the process mean shifts upwards, and we are now sampling from the solid distribution. The change is soon picked up in the X-bar chart.

12.4   Individuals Chart

In order to establish an individuals control chart, we need to collect at least 20 measurements from the process in which we are interested. The *three-sigma* limits are calculated from the standard deviation of the data to give the control limits as:

$$\text{LCL} = \text{mean} - (3 \times \text{standard deviation})$$

and

$$\text{UCL} = \text{mean} + (3 \times \text{standard deviation})$$

The control chart is obtained by drawing these limits and the centre line (the mean) on a run chart of the individual values. Care has to be taken when calculating the standard deviation. If special causes have occurred during the data collection period, the size of the standard deviation can be affected and, hence, the width of the control limits. For instance, in the debt collection problem in Section 1.3, the January figures are lower than other monthly data. A reason for this has been established, namely that (in the southern hemisphere) January is the month when most businesses close down for the summer vacation. With these special causes present the standard deviation will be higher than it would be otherwise. In this case, it would probably be sensible to calculate the standard deviation with the January figures omitted. Other methods of calculating the variability based on a *moving range* have been suggested; see, for example, Joiner's *Fourth Generation Management* (p. 234).

Eldon Cycle Cranks

Jim Boardman, the sales manager of Eldon Cycles, has doubts about whether their new 75-mm cycle cranks are continuing to sell since their major competitor brought their own product onto the market last July. Consequently, Jim has collected data on the number of cranks sold per month for the last 3 years. He has calculated the control limits and constructed the control chart given in Figure 12.8.

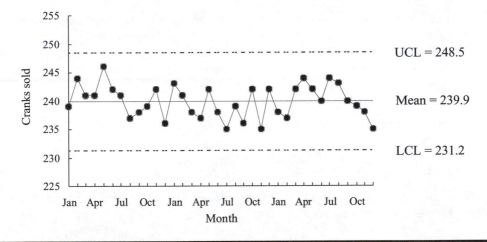

Figure 12.8   Control Chart for Cycle Crank Sales

Q9. What should Jim conclude from his analysis?

Q10. What should Jim do next?

12.5   Attribute Charts

So far we have considered control charts for continuous or interval data (X-bar and R charts, and individuals charts). These are known as *variables control charts*. We shall now consider two other types of control charts, known as *attribute control charts*, for data that are either proportions or counts. The charts are

- the p-chart for the proportion of defective items. For example:
  - ➤ the proportion of ball point pens in a box of 20 that do not write on purchase.
  - ➤ the proportion of plastic milk containers that leak.
- the c-chart for the counts of defects on a unit. For example:
  - ➤ the number of paintwork blemishes on a finished car.
  - ➤ the number of typing errors per page.

In calculating the three-sigma limits for these two control charts, we shall use results for proportion data given in Section 5.5 and for counts data in Section 5.6.

The p-Chart

The steps to follow in the construction of the p-chart are

1. Collect at least 20 subgroups of n items each, and calculate the proportion of defective items in each subgroup.
2. Calculate the average proportion of defective items $\bar{p}$.
3. Plot the proportion of defective items in each subgroup and the centre line $(\bar{p})$ on a run chart.
4. Calculate the *three-sigma* control limits, and add them to the chart. The limits are given by

$$\text{LCL} = \bar{p} - 3\sqrt{\frac{\bar{p}(1-\bar{p})}{n}} \qquad \text{UCL} = \bar{p} + 3\sqrt{\frac{\bar{p}(1-\bar{p})}{n}}$$

5. Find any out-of-control points and remove them from the chart if a special cause can be found. Recalculate the centre line and the control limits.

Dotoya Cars Ltd.

Dotoya Cars Ltd. assembles cars for sale in Australia. Recently they have been having problems with the assembly of rear seats, with some of the laminated plastic coming away from the seams. They examined this problem over a period of 30 weeks. Each week they took a random sample of 40 seats and recorded the number with defect seams. The results are given in Figure 12.9, and plotted in Figure 12.10.

Q11. Calculate the control limits and draw them on the graph in Figure 12.10.
Q12. What can you say about this process?

Week	Number of defective seats	p	Week	Number of defective seats	p	Week	Number of defective seats	p
1	3	0.075	11	9	0.225	21	4	0.100
2	6	0.150	12	4	0.100	22	7	0.175
3	3	0.075	13	6	0.150	23	3	0.075
4	4	0.100	14	3	0.075	24	6	0.150
5	2	0.050	15	2	0.050	25	8	0.200
6	4	0.100	16	3	0.075	26	5	0.125
7	3	0.075	17	6	0.150	27	2	0.050
8	7	0.175	18	3	0.075	28	4	0.100
9	4	0.100	19	6	0.150	29	7	0.175
10	4	0.100	20	6	0.150	30	4	0.100
Total	40			48			50	

Figure 12.9   Numbers of Defective Seats

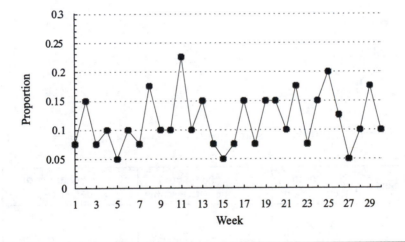

Figure 12.10   Proportion of Defective Seats

 The c-Chart

The steps to follow in the construction of the c-chart are

1. Obtain at least 20 similar inspection units, and count the number of defects on each unit. Note that the inspection unit has to be large enough to produce observable effects.
2. Calculate the average number of defects $\bar{c}$.
3. Plot the number of defects per inspection unit and the centre line ($\bar{c}$) on a run chart.

4. Calculate the three-sigma limits, and add them to the chart. The limits are given by

$$\text{LCL} = \bar{c} - 3\sqrt{\bar{c}} \qquad \text{UCL} = \bar{c} + 3\sqrt{\bar{c}}$$

5. Find any out-of-control points and remove them from the chart if a special cause can be found. Recalculate the centre line and the control limits.

Dotoya Cars Ltd.

After consulting a statistician, Dotoya Cars Ltd. decided to look at their defective car seat problem again. The statistician advised them it would be better to set up a c-chart, where they count the number of seams that are defective on a seat, rather than classifying seats as defective or nondefective.

Q13. Why has the statistician given this advice? What are the merits of this approach, compared to the p-chart approach?

Data collected from 30 seats are given in Figure 12.11.

Seat	Number of defective seams	Seat	Number of defective seams	Seat	Number of defective seams
1	7	11	8	21	6
2	5	12	7	22	6
3	0	13	4	23	4
4	8	14	6	24	7
5	3	15	2	25	7
6	6	16	7	26	3
7	5	17	5	27	3
8	4	18	5	28	3
9	5	19	3	29	3
10	5	20	5	30	3

Figure 12.11   Numbers of Defective Seams

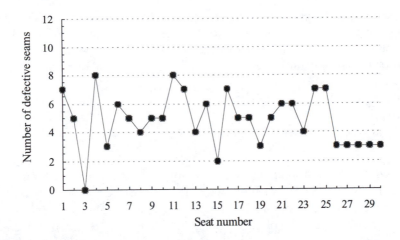

Figure 12.12   Number of Defective Seams on Car Seats

Q14. Construct the control limits for this example and plot them on the chart in Figure 12.12.

Q15. What can you say about this process?

12.6   Exercises

1. When a point is above the upper control limit on a range chart, then most likely
   a. the process mean has shifted upwards
   b. the process variation has decreased
   c. the process variation has increased
   d. nothing has changed
   e. the process mean has shifted downwards

2. Control charts are used to distinguish between
   a. defectives and nondefectives
   b. common cause and special cause variation
   c. good and bad work practices

3. Control limits are derived from
   a. engineering knowledge
   b. process data
   c. management directives
   d. machine specifications
   e. none of the above

4. The manager of each branch of Bonzo Burgers keeps a record of all customer complaints. The number of customer complaints each week is part of the data sent to head office. If head office wants to put in place a control chart to track the number of complaints, which of the following would you advise?

 a. a p-chart

 b. a c-chart

 c. X-bar and R charts

 d. an individuals chart

5. G&S Whiteware Manufacturers are having difficulties in machining spindles for their new Soft Suzie washing machine. They decide to set up control charts for the width of the spindle by sampling 4 spindles every hour, for the next 2 days or so. The widths of the spindles in millimetres are given in Figure 12.13,

G&S Whiteware

	A	B	C	D	E	F	G	H
1	Subgroup	Item1	Item2	Item3	Item4	Mean	Range	Notes
2	1	10.3	11.5	9.8	12.3	11.0	2.5	
3	2	9.9	8.7	10.1	11.2	10.0	2.5	GH absent
4	3	13.2	12.9	13.4	13.3	13.2	0.5	
5	4	12.6	12.3	11.9	12.2	12.3	0.7	
6	5	10.0	12.8	12.8	12.4	12.0	2.8	
7	6	10.5	13.1	13.2	13.1	12.5	2.7	
8	7	12.3	11.1	13.1	15.9	13.1	4.8	New operator
9	8	11.9	10.1	13.0	13.4	12.1	3.3	
10	9	11.8	9.9	12.9	12.2	11.7	3.0	
11	10	11.7	9.8	13.3	11.1	11.5	3.5	
12	11	10.3	10.6	10.5	13.8	11.3	3.5	
13	12	10.2	12.4	14.2	12.2	12.3	4.0	
14	13	12.6	13.9	10.7	13.1	12.6	3.2	
15	14	10.8	13.2	14.2	12.0	12.6	3.4	
16	15	9.8	11.1	12.2	13.0	11.5	3.2	
17	16	12.3	12.8	13.2	13.5	13.0	1.2	
18	17	11.0	13.2	14.3	12.6	12.8	3.3	
19	18	9.0	14.0	12.3	12.3	11.9	5.0	Oil ran out
20	19	15.0	12.3	13.4	12.9	13.4	2.7	
21	20	12.0	12.5	12.2	14.3	12.8	2.3	
22	21	12.5	12.4	12.5	12.6	12.5	0.2	
23	22	9.5	12.6	13.2	13.2	12.1	3.7	
24	23	11.3	11.6	11.9	12.3	11.8	1.0	
25	24	14.3	12.4	11.6	12.5	12.7	2.7	
26	25	9.3	12.7	10.4	12.3	11.2	3.4	
27						12.14	2.76	

Sheet1 / Sheet2 / Sheet3 /

Figure 12.13   Spindles Data

together with subgroup means, ranges, and overall means (in F27 and G27).

    a. Calculate the control limits, and set up control charts.

    b. What conclusions do you draw from your control charts?

6. Consider the Nuva Plastics data given in the Excel spreadsheet in Figure 12.3. No special causes showed up on the R chart in Figure 12.4 or the X-bar chart in Figure 12.5. The process that produced these data is, therefore, a stable process. Or is it?

    a. Look closely at the data in each subgroup. What do you notice?

    b. Why do you think it happened?

    c. What do you suggest as a remedy?

7. Helen Cassells was reviewing the figures for last year's video camera sales. One of the models, the Z20A, had been selling well at the beginning of the year but now looks to be in decline. She called Neville Watson, brand manager in charge of this model's sales, to see what had been happening.

"Neville, have you seen the latest figures on the Z20A? Since you employed that new salesperson at the end of October, there's been a downturn in sales."

"Yes, Helen, I'd noticed that too, but I think we should look at the monthly figures for the last 2 years before we jump to any conclusions."

"OK, Neville, but can you get back to me about this before the end of the month with a recommendation about what we should do to stop a further decline in the sales of the Z20A?"

Neville collected the sales figures for the last 2 years, and they are given in Figure 12.14. The mean sales is 17 ($000) with a standard deviation of 1.10 ($000).

Month 1993	Jan	Feb	Mar	Apr	May	Jun	Jul	Aug	Sep	Oct	Nov	Dec
Sales ($000)	15.4	15.6	17.2	16.8	16.3	16.6	17.3	18.0	17.2	16.5	17.5	19.2
Month 1994	Jan	Feb	Mar	Apr	May	Jun	Jul	Aug	Sep	Oct	Nov	Dec
Sales ($000)	18.2	17.3	17.4	16.3	15.4	17.2	17.3	15.5	18.9	19.1	16.4	16.1

Figure 12.14 Sales of the Z20A Model

    a. Calculate the control limits for these sales and draw them on a control chart.

    b. What conclusions do you draw from the control chart?

    c. What should Neville do next?

8. In Question 9 in Section 11.13 a policy to control costs was discussed. In order to examine this policy more closely, the cost variances, expressed as a percentage over or under budget, on the last 20 construction projects were obtained. They are given in Figure 12.15.

Project	A	B	C	D	E	F	G	H	I	J
Cost Variance	-12.5	0.1	-6.1	10.1	10.1	9.5	-12.5	1.3	1.3	32.0
Project	K	L	M	N	O	P	Q	R	S	T
Cost Variance	-0.7	7.8	0.7	-5.7	3.5	-9.1	-5.2	9.1	0.7	-5.9

Figure 12.15   Project Cost Variances (%)

The mean and standard deviation are 1.425% and 10.17%, respectively.
a. Calculate the upper and lower control limits, and plot them on a control chart.
b. Review your answer to Question 9 in Section 11.13, namely, "What are your reactions to this policy?"

9. Elaine is a line supervisor responsible for filling cans with milk powder at a local dairy factory. She has been having problems with getting the correct amount of powder in the cans. She has to ensure that she does not underfill the 2-kg cans, but at the same time the company does not want to give too much powder away. There are two settings on the can filling machine, a high setting and a low setting. Elaine is collecting data on this process, and when she gets a result that she regards as too high she resets the machine to its low setting, but if it is too low she resets it to its high setting. Figure 12.16 gives the control chart constructed from the data Elaine collected.
   a. Is there any evidence of special causes in this control chart?
   b. What do you think has happened here?

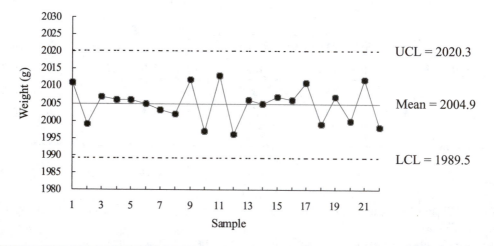

Figure 12.16   Weight of Cans of Milk Powder

10. A plastics company is looking into the variability of the percentage content of recycled plastic used for moulding plastic buckets. Customers have been returning buckets that crack too easily when dropped. The company believes this is due to the recycled plastic content of the buckets, which should never be more than 40%. Figure 12.17 gives a run chart of the percentage of recycled

plastic measured from 24 samples taken from the feed hopper at 10-min intervals over 1 shift.

a. Comment on the state of the process.

b. What steps would you take in investigating this problem?

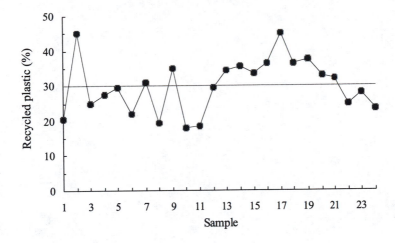

Figure 12.17   Percentage Recycled Plastic

Chapter 13

Improvement Strategies

13.1 Monitoring the Process

Once appropriate control charts have been established, they can be used for monitoring by adding further points to the charts for each subsequent period as additional data are collected. This task is usually the responsibility of operators and supervisors, so they can get immediate feedback.

Prompt reaction to out-of-control points is essential for successful process improvement because:

- It is much easier to find special causes on the spot rather than reconstructing the circumstances later, if at all possible.
- Charts are very conservative. When a signal occurs it is almost certain a special cause is acting, and there is an opportunity to better understand and improve the process.
- Lack of reaction causes all involved to doubt the usefulness of the control charts.
- It is better to have no chart at all than to ignore the signals from a chart in place.

Resources must be made available to search for special causes once a signal occurs. The project team should review the charting process periodically. If no signals are occurring it is wise to confirm that the data are being collected according to plan. Consideration can be given to changing the nature of the subgroups, the frequency of sampling, and so on. If the variation in the process has been reduced by removing special causes, the control limits should be revised on the basis of recent data.

Dotoya Cars Ltd.

Consider again the defective seat problem discussed in Section 12.5. Data collected on 30 seats were used to establish the chart. The resulting c-chart is given in Figure 13.1.

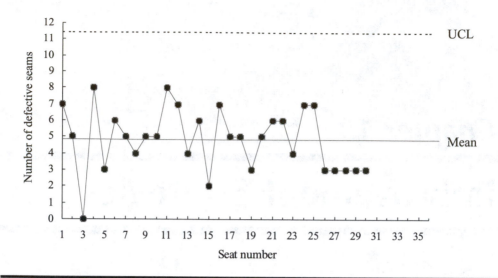

Figure 13.1   c-Chart for Defective Seams

Q1. What can you say about the stability of the process? Is the process in statistical control?

The chart is now used to monitor the process. Data collected from the next 7 seats had 3, 3, 2, 3 1, 0, and 3 defective seams.

Q2. Plot these data on the control chart. What conclusions do you now draw?

Q3. What would you suggest Dotoya does next?

13.2   Sensitivity of Control Charts

A forest products company constructed a control chart for one of their processes, but when they used the control chart for monitoring the process they discovered that the chart signaled special causes every hour. They could not afford to stop the process because of lost production. Neither could they provide resources and time for operators to investigate the process every hour. They concluded that the use of control charts caused more problems than they solved; so they were discontinued.

What they failed to understand was the sensitivity of the chart. The chart was signaling too often for the resources that were available. Charts can be made less, or more, sensitive by changing the frequency of sampling and/or the size of the subgroup sample. Generally, the more frequently a process is sampled, or the larger the size of the subgroup, or both, the more sensitive the chart.

Consider again the Nuva Plastics example in Section 12.2. The line manager, Ellen Vagner, took samples of five baskets at five different times of the day.

> **Q4.** *In establishing the chart do you think the subgroup size and frequency of sampling is appropriate? What can you say about the sensitivity of this chart?*
>
> **Q5.** *What are the possible dangers of sampling five times a day?*
>
> **Q6.** *How is the sensitivity of the chart affected by increasing the frequency of sampling to five per hour?*

13.3 Specifications

One dimension of quality is *conformance to specifications*. The assumption is that if all parts of a product conform to specifications, the entire product will be of good quality.

In the manufacturing and process industries, specifications are set for the most critical characteristics. For example, a dairy company may aim to fill 2-kg cans of milk powder with no more than 2.01 kg of powder. The specification limits are, therefore, 2 to 2.01 kg. Any cans containing less than the lower specification limit (LSL) would violate their legal requirement, while filling with more than the upper specification limit (USL) would result in the dairy company giving away too much powder.

Again, suppose we are making electric motors. Our requirements are that the main shaft is round and is 25 mm in diameter. The target is 25 mm, but not all shafts will be 25 mm in diameter as some variability is inevitable. As long as they are not too much off target, they can be used. Specification limits might therefore be set at 25 ± 0.1 mm.

The same concepts apply with business processes, and in the service industries. A freight company might set specification limits on the time it takes to deliver a parcel. A delivery that takes longer than a certain time (USL) may involve some penalty. For example, one parcel delivery company guarantees delivery within

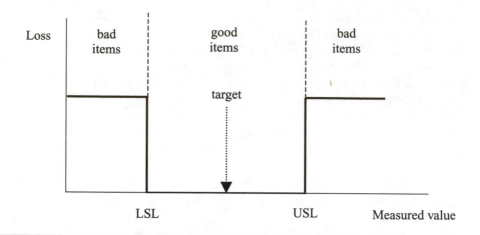

Figure 13.2   Conventional View of Specifications

24 hours, or you get your money back. A telephone company may say that they are providing a good service if your telephone fault is rectified within 8 hours. The finance department may be content if they pay travelling expenses within 2 weeks. An airline may have a specification that all baggage must reach the carousel within 20 min of the plane arriving.

The conventional view of specifications is an all or nothing situation. Items are either good or bad. The baggage either arrives within the specified time limits or it does not. This view is depicted in Figure 13.2.

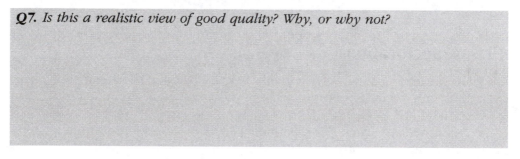

Q7. Is this a realistic view of good quality? Why, or why not?

The conventional view that conformance to specification equates to good quality is oversimplistic. There is little interest in whether the target has been met exactly. Variability *within* specification is, for the most part, ignored. A further drawback is the problem known as *tolerance stack-up*. When two or more parts are to be matched together the fit might be poor if one part is at the lower end of its specification and the matching part at the upper end. Even if the parts are matched initially, the link between them may wear more quickly than one made from parts whose targets have been matched exactly. Repairing products at a later time can also cause problems in finding matching parts.

A more realistic view is given by the loss function depicted in Figure 13.3. This way of looking at specifications is due to the Japanese engineer Genichi Taguchi. The ideal is at the target and any movement away from that causes increasing losses. Losses near the target value may be negligible, but as you move further away they become greater. The loss function is *not* a step function.

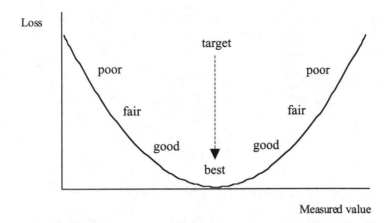

Figure 13.3   Taguchi's Loss Function

We have ever-increasing losses with departure in either direction from the target value. Neither this function, nor the one in Figure 13.2, has to be symmetric. It may even be one-sided, as with the telephone company where the ideal is to fix the fault immediately.

Consider the following situation where we have the choice between two suppliers A and B for some component. The distribution of the critical dimension of the component is shown for both suppliers in Figure 13.4, together with the specification limits.

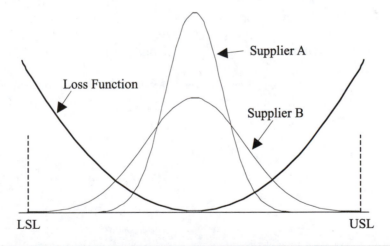

Figure 13.4   Component Distributions for Suppliers A and B

Q8. Both suppliers have all their components within the specification limits. Which supplier would you choose, and why?

Q9. *Suppose Figure 13.4 referred to two airlines delivering baggage to the carousel. Which airline would you favour, and why? (Here the loss function is probably not symmetric.)*

13.4   Specification and Control Limits

We have now considered two types of limits, namely *control limits* and *specification limits*. They are often confused in practice, but it is important not to do so.

Control limits are to do with the performance of the process. They tell us what the process is doing right now. They are calculated from the data collected on the process. They are the *voice of the process*. We use control charts to help us distinguish between special and common causes of variation. The aim is to eliminate the special causes so that we have a stable and predictable process. It does not necessarily mean that we are happy with the process, or that it meets our needs.

Specification limits reflect what we desire from the process. They are limits ultimately set by the customer, and are the *voice of the customer*. They are chosen by the process engineer to reflect what the customer wants. For example, if the customer wants the door on the refrigerator to close properly and tightly each time, specification limits are set on factors such as the size of the door, hinge mechanism, and frame dimension, in order to achieve that requirement. They are not determined by the data.

It is important **not** to draw specification limits on a control chart. It is a dangerous practice that can badly mislead you.

Figure 13.5 gives the X-bar chart for the Nuva Plastics example of Section 12.2. Suppose the specification limits are 113 ± 0.5 g. Draw them on the X-bar chart (never do this again!).

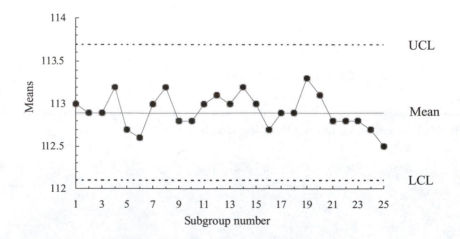

Figure 13.5   Nuva Plastics X-bar Chart

Q10. Are all the data within specification limits?

Q11. In what way can you be misled by what you have done?

13.5 Process Capability

Suppose we have a process that is in statistical control. Does this mean that the process meets customer needs, as set by the specification limits? Is our entire production within the specification limits, or do we have to reject some items because they fall outside these limits. In other words, is our process *capable*?

A process can either be in statistical control or not. The product can either meet the specifications or not. Hence, when considering the capability of a process, there are four states to be considered.

 ### *Ideal State*

The ideal state is when the process is in statistical control and all of the output is within the specification limits. It means that all the output is suitable for the intended purpose, and as long as it stays in control the process will produce consistent output hour after hour, day after day, week after week. The aim is to stay in this state. Control charts are used to signal problems before they are severe enough for the process to become incapable. Even in this state there may be reasons for further improving the process. Lower costs and greater productivity can be achieved from even more uniform output, as shown by Taguchi's loss function in Figure 13.3.

 ### *Threshold State*

If the process is stable but some of the output is outside the specification limits, then we are in what is called the threshold state. A process in this state is judged incapable. For a process that is stable but incapable, the process may not be centred on target, may be too variable, or maybe both.

Different remedies will be needed to resolve the two different sources of incapability.

- The process can often be centred on the target by changing the operating parameters. Experimentation often reveals how to make these adjustments.
- The reduction of excessive variability requires removal of some of the common causes of variation, and is usually more difficult.

Control charts should be used to ensure that the process is maintained in statistical control, and also as a device to monitor improvement.

Brink of Chaos

We are in the brink of chaos state if the process is unstable, but all of the output is within specification. But what is wrong with that? Surely customer needs are being met.

It is not possible to say anything about the capability of a process that is unstable. Since special causes are present, we no longer have a predictable process. Although, in this state, the product right now is all within specification, it is likely to wander off and produce output out of specification. The problem is that special causes are determining what the process is producing. Quality and conformance can change at random, thus leading to chaos. It is necessary to eliminate these special causes.

Chaos

Chaos results if the process is out of control, and is producing nonconforming products. Again the process is dominated by special causes. The only way out of this state, into the ideal and threshold states, is first to eliminate special causes. Control charts are again invaluable as a tool to assess whether special causes are being eliminated.

Airport Immigration

An international airport has to maintain processing standards set by the government by ensuring that 95% of passengers arriving on a flight are processed through immigration within 45 min of landing. Data collected for all flights landing over 2 days are given in the stem and leaf diagram in Figure 13.6. Each observation gives the time (to the nearest minute) for 95% of passengers from a flight to be cleared through immigration. The data were obtained from the computer system used to record arriving passengers. Mark in the specification limit on the stem and leaf diagram.

```
2 | 2
2 |
2 | 7
2 | 8 8 9 9
3 | 0 0 0 0 0 1 1 1 1
3 | 2 2 3 3 3 3 3
3 | 4 4 4 4 4 4 4 5 5 5 5 5
3 | 6 6 6 6 6 6 6 6 6 6 7 7 7 7 7 7
3 | 8 8 8 8 8 8 8 9 9 9 9
4 | 0 0 0 0 0 0 0 0 0 0 1 1 1 1 1
4 | 2 2 2 2 3
4 | 4 4 4 4 5
4 | 6
4 |
5 | 0 0 1
```

Figure 13.6   Immigration Processing Times

Q12. *What percentage of flights is not meeting the standard?*

Q13. *What type of control chart is appropriate for these data?*

The control chart is given in Figure 13.7.

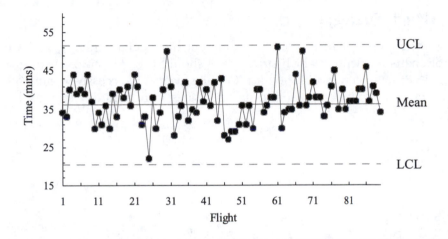

Figure 13.7   Control Chart for Immigration Processing Times

Q14. *What is the capability state of this process?*

Q15. *What is the first step to take in improving the process?*

13.6   Improvement Strategies

We have seen that control charts can be used to identify whether variation arises from common or special causes. Further, for a process in one of the two states of chaos, it is important to first eliminate special causes. In the ideal and threshold states, further improvement will only occur if some of the common causes are identified and eliminated. Different improvement strategies are required for processes that exhibit common cause or special cause variation.

Sometimes special causes represent the outcomes we seek, for example, increased sales as a result of a TV advertising campaign. However, if the special causes have undesirable consequences, then they need to be eliminated. This means finding out what was different at the time they occurred in order to prevent them from reappearing. Eliminating special causes is the first priority. They are usually easier to detect and fix, and tend to mask the common cause variation.

Remember the 85/15 rule though, where most problems are due to *common causes*. Common cause problems are harder to detect, involve changes to the system, and invariably require management action. However, once we have succeeded in dealing with the special causes, further improvements can only be made by developing strategies to reduce common cause variation.

Which Strategy?

Fixing each type of cause requires a different managerial strategy, and we use control charts to identify which strategy we should use. A control chart, given in Figure 13.8, has been established for TV sales data on successive Saturdays.

The chart is now being used to monitor sales from Saturday to Saturday.

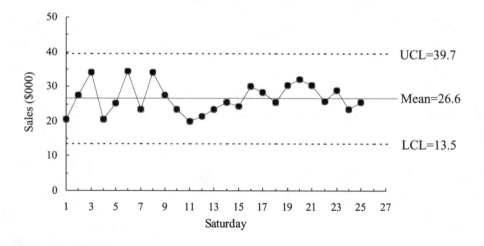

Figure 13.8   TV Sales

> **Q16.** Which improvement strategy would you use if TV sales were $47,000 on the 26th Saturday?
>
> **Q17.** What if sales were $31,000?

13.7   Special Cause Strategies

Steve arrived in the office that morning to discover that Kaplan's order had not been delivered on time. For the last 3 years, the monthly order had always been on time. What had changed?

Here we probably have a special cause, with an opportunity to improve delivery performance. Not all special causes are so obvious and we need control charts to spot them, but if we miss the opportunity of finding out what is different now the problem is likely to recur.

Steve phoned Kaplan and spoke to their purchase manager, and explained that the order would be a day late. He then set about looking into the problem. Upon investigation Steve discovered that Kaplan's order had been late because they had to rerun the whole batch of boxes again because somebody had printed Kaplan's new telephone number upside down on the outside of the box. Further investigation revealed that this had happened on several orders, and was caused by a recent change in local telephone numbers.

A common mistake is to fix the problem by adding extra complexity to the process—in this case, perhaps by adding an inspection stage before the batch of boxes is printed. Instead they redesigned all printing "slugs" by changing them from clean oblong shapes, to oblong shapes with a protruding notch, as depicted in Figure 13.9. In this way, they could not be put into the printing machine upside down.

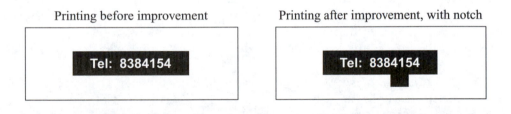

Figure 13.9   Redesigned Printing Slugs

The example illustrates the important stages in dealing with special cause problems.

1. Get timely data, so special causes are signaled quickly. A speedy reaction to a special cause problem is very important. We must therefore get data immediately so that we can investigate the problem when the trail is hot. For this reason it is best to place the control chart near the process.
2. Put in place an immediate remedy to contain any damage. This is "fire-fighting" but it is necessary, for example, to take action to soothe an angry customer. Fire-fighting is not acceptable if we were to stop there.
3. Find out what was different about the special cause. Has something changed? Is there perhaps a new operator, or a new competitor, or a change in procedure?
4. Develop a long-term fix. We should have learned something in the search for the cause. What can we do to stop its recurrence, or if the results were good, maintain this level of performance? This could mean changing some higher-level system, which often brings benefits beyond the immediate problem. In Steve's case he changed the printing procedure and this prevented upside down printing on all orders, not just Kaplan's and not just telephone numbers.

Q18. *You have been getting A's all year in assignments and tests. Suddenly you get a C. How would you apply special cause strategies to this problem?*

13.8   Common Cause Strategies

A process with only common cause variation is stable and predictable. However, it may not be meeting customer needs. For example, the beads process is a stable and predictable process, but we are producing too many red beads (defects).

Improving a process with only common causes of variation present needs a different approach to that of eliminating special causes. Trying to find out what was different between this data value and the others, as we would with a special cause strategy, will not necessarily yield any answers. All the data needs to be examined.

Common causes are usually difficult to discover. They do not suddenly present themselves like special causes. They are there all the time and need to be found by analysing all the data. With common causes, quick fixes rarely work. What we need are strategies that help us understand the workings of hidden causes present all the time in the system. There are three major strategies we can adopt.

- *Root Cause Analysis.* We must be able to identify the major potential causes that lead to the problem. We need to dig down to find the root causes. This means not just accepting the first explanation of what is happening, but to keep on questioning and probing until the real causes have been determined. The cause and effect, or fishbone, diagram is a useful tool for displaying the potential causes. Recall how, in the debt collection problem in Chapter 2, the causes were identified and verified, and solutions found that reduced variation and the level of outstanding debt.
- *Stratification.* We can look at data that has been collected from the process by sorting it into groups or categories based on different factors. In this way we can seek patterns in the way the data cluster or do not cluster. For example, TV sales data varies from month to month. We could stratify it by region, product line, day, and so on, and look for patterns in each stratum. We can collect data on different parts of the process. For example, in a manufacturing process made up of several stages (production, storage, dispatch, and transportation), delivery dates have been falling behind schedule. Collecting and analysing data from each stage may help discover common causes of late delivery.
- *Experimentation.* This involves making carefully planned changes to a process, analysing the results, and observing any effects. For example, in the production of corrugated cardboard, it is known that the temperature

and pressure of the rolling process, and the type of glues used, affect board thickness. We might run an experiment with different temperatures, pressures, and glues to discover the best settings. Experimentation is usually expensive and can often disrupt the process, but it does allow us to find causal relationships between inputs and outputs.

The aim is to localise common cause variation by pinpointing it at its source. Once we know that, we can develop measures to counteract the common cause variation by making *fundamental changes* in the system.

For example, consider the extrusion process example in Section 11.4. Bernard's job was to make sure that the extruder produced aluminum tubing of 50-mm diameter. Various strategies were considered to help Bernard to try to meet his customers' needs. He left the process alone, and he tried to adjust the process. Nothing he tried worked.

> **Q19.** *What strategies should Bernard have adopted in order to improve the results?*

As before, we can use the quincunx to mimic Bernard's process. In Section 11.4 we used the quincunx with the pin board shown in Figure 13.10. Each row of pins introduces variability into the results, since in each row a ball can go to either side of the pin. Each row represents a common cause.

Suppose we were able to eliminate half the common causes. With the quincunx we can mimic this by using the pin board shown in Figure 13.11. Note that the last five rows are in line, so that when a ball reaches the sixth row it will continue straight through.

The first strategy in Section 11.4 involved leaving the funnel above the 50-mm slot at all times. We repeated this by dropping 500 balls using each of the two pin boards in turn. A bar chart of the results is given in Figure 13.12.

Figure 13.10   Full Pin Board

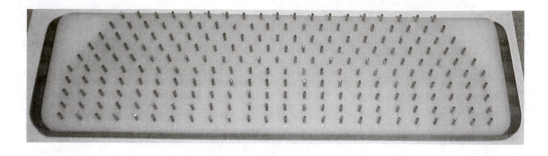

Figure 13.11   Half Pin Board

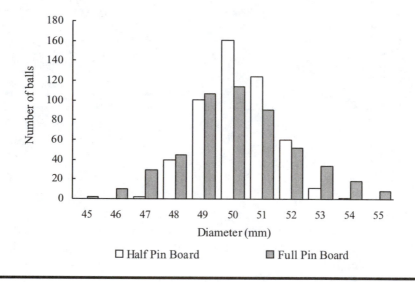

Figure 13.12   Results from Each Pin Board

Q20. What do you conclude from this bar chart?

Cape Corp Debt Collection Agency

Midi Vaartagen, the manager of Cape Corp Debt Collection Agency, was reviewing progress on the unpaid TV licences contract. Over the last few months it looked as if there had been a drop in recovered TV licence debts. Midi decided not to jump to any conclusions. The last time she thought there was a drop in the recovery of debts, it turned out to be a temporary downturn and she had upset

1997	Jan	Feb	Mar	Apr	May	Jun	Jul	Aug	Sep	Oct	Nov	Dec
%	25.6	23.4	25.8	26.2	27.3	24.5	25.2	23.5	27.1	26.1	27.2	28.0
1998	Jan	Feb	Mar	Apr	May	Jun	Jul	Aug	Sep	Oct	Nov	Dec
%	26.3	25.3	24.7	24.8	25.0	24.6	24.2	24.0	23.9	24.4	23.1	24.2

Figure 13.13 Percentage Recovered Debts

quite a few of her staff when she called them in to explain what was happening. So this time she thought she would follow some advice on collecting data that she vaguely remembered from the first year at university when she was studying for her management degree. Figure 13.13 gives data on the percentage of debts they had managed to recover for the TV licensing authority over the last 24 months. The mean of this data is 25.18% and the standard deviation 1.34%.

Q21. What type of control chart should we use for these data?

Q22. Calculate the control limits and plot them on the run chart in Figure 13.14.

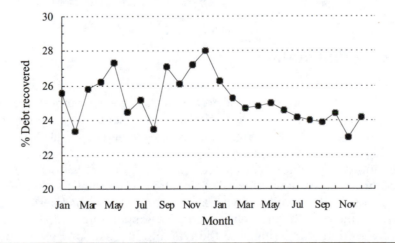

Figure 13.14 TV Licence Debt Collection

Midi thought about possible actions that she could take, and came up with the following five options.

1. It looks as if the campaign on one of the pirate radio stations may be encouraging people not to pay their TV licences. Contact the broadcasting minister about this.
2. Dig out further data on TV licence debt collection and categorise them according to employee, department, part of country, etc. to see if any patterns emerge.
3. Figure out what was different during the last few months compared to other months. Were there more debts to investigate? Were there new employees in the department? Was there more or less than normal advertising on TV?
4. Look at all the debts that had to be processed in the past few months and categorise the causes of the problems. Look for patterns.
5. Have people working on each major stage in the debt collection process collect data for several weeks on how many, and what kinds, of errors occur in each stage.

Q23. What should Midi do?

Q24. Suppose Midi decides to use a control chart to monitor the progress of TV licence debt collection. What should be her next step?

13.9   Experimentation

When it is difficult to discover the reasons for common cause variation from data that already exist, or from data collected from the process running under normal conditions, we may decide to carry out an experiment on the process. This is done by making planned changes in process variables, collecting data, and studying the effects of the changes.

A machine that produces corrugated cardboard (a corrugator) was improved using a planned experiment. One of the qualities important for corrugated cardboard is thickness. It is well known that variation in the thickness of the board depends on numerous machine settings, such as temperature, pressures of different rollers, the speed of processing, and the type of glue. The process is so complex that data collected on board thickness had yielded no solution to the problem of excessive variation in thickness; so, an experiment was undertaken. This involved setting the corrugator at different temperatures, roller pressures, and speeds in a planned way to discover the causes of the variation.

The operations manager of a plastics company was concerned by the variability in the shrinkage of product obtained from an injection moulding machine. She decided to carry out a planned experiment to try to discover the causes of the excessive variation. Her experiment involved setting the levels of a number of variables she considered likely to affect output. These included cycle time, mould temperature, holding pressure, injection speed, holding time, and gate size.

Other examples are planned experiments to determine the causes of variation in the moisture content in particle board, or in the temperature in a furnace, or in the thickness of cling film used for wrapping.

Many issues and concepts are involved in planning and carrying out useful industrial experiments. We shall demonstrate some of these with the helicopter experiment.

 ## *The Helicopter Experiment*

The aim of the helicopter experiment is to show you the power of statistical experimentation in uncovering hidden factors that influence the performance of a process. In this demonstration we look at how we can improve the design of a paper helicopter. All of the ideas extend to real industrial experiments, like that of the corrugator or the injection-moulding machine. A picture of a helicopter is given in Figure 13.15.

> **Q25.** *How will we measure the performance of our paper helicopter? How would we know if we had made a good one?*
>
> **Q26.** *There are many things we can vary in the design of the helicopter. Make a list of the factors you think are likely to influence the helicopter's performance.*
>
> **Q27.** *Each of the factors listed can take many different values, or levels. What levels should we use for each?*

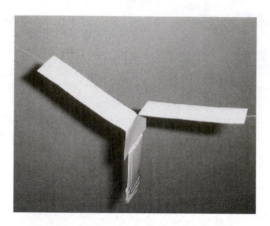

Figure 13.15   A Paper Helicopter

    Designing the Helicopter

Suppose we have chosen the factors and levels shown in Figure 13.16.

> **Q28.** *How many different paper helicopters can we make?*
>
> **Q29.** *We could make all the helicopters, but it would take a lot of time to carry out the experiment. It would also be expensive and time consuming. How else could we run the experiment?*

	Factors	Levels	
1	Type of paper	thick	thin
2	Rotor width	2cm	4cm
3	Rotor length	10cm	12cm
4	Weight	no clips	2 clips
5	Type of shaft	fold	cut
6	Fold on bottom	no	yes

Figure 13.16   Helicopter Factors and Levels

Helicopter Experiment Design

The experiment is run with eight different helicopters involving six factors each at two levels. In Figure 13.17 the run number is given in the first column, and the *design* of the helicopter in the next six columns. The last two columns will be used to record the results of the experiment.

> **Q30.** *What patterns can you see in the levels of the factors in the design in Figure 13.17? (Hint: Look carefully at each column and each pair of columns.)*
>
> **Q31.** *Should we run the experiment in the order given in Figure 13.17? If not, how do we decide the run order?*

Run	Type of paper	Rotor width	Rotor length	Weight (Paper clips)	Type of shaft	Fold on bottom	Time	Quality
1	thick	2 cm	10 cm	0	cut	yes		
2	thin	2 cm	10 cm	2	fold	yes		
3	thick	4 cm	10 cm	2	cut	no		
4	thin	4 cm	10 cm	0	fold	no		
5	thick	2 cm	12 cm	0	fold	no		
6	thin	2 cm	12 cm	2	cut	no		
7	thick	4 cm	12 cm	2	fold	yes		
8	thin	4 cm	12 cm	0	cut	yes		

Figure 13.17   Design of the Helicopter Experiment

 ### *Plotting the Data*

Two responses are obtained when we drop each helicopter, and these are to be recorded in the last two columns in Figure 13.17. If you are not running the experiment, copy the data given in Figure 13.21 into Figure 13.17. The first response is the time (in seconds) for the helicopter to land on the floor. The second is a measure of the quality of the flight, on a scale from 0 to 5. A beautifully spinning, slowly descending helicopter is a 5, and one that flops miserably is a 0.

Totals	Type of paper		Rotor width		Rotor length		Weight (No. of clips)		Type of shaft		Fold on bottom	
	thick	thin	2cm	4cm	10cm	12cm	0	2	fold	cut	yes	no
Time												
Quality												

Figure 13.18   Total Scores for the Helicopter Experiment

Having collected the data, we can calculate the total scores for each factor level. For example, the time and quality score for a 10-cm rotor length are the totals of rows 1 to 4 in Figure 13.17, whereas the values for 2 paper clips are obtained by summing rows 2, 3, 6, and 7. These total scores can be entered into Figure 13.18.

The total scores for time can now be plotted in Figure 13.19, and the total scores for quality in Figure 3.20. The scores for the two levels for each factor are then joined to show the change caused by varying that factor.

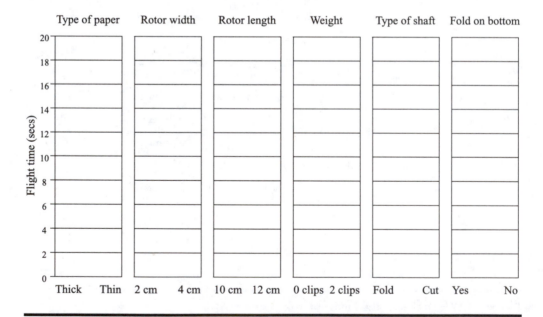

Figure 13.19   Plot of Helicopter Flight Time

✍ *Finding the Best Helicopter*

The plots in Figures 13.19 and 13.20 will help you determine which factors might influence the choice of the best helicopter. An important factor will have a large difference between the results at its two levels, depicted by a steep line in your plot from the one level to the other level. Where the choice of level of a factor is not important, the lines will be fairly horizontal.

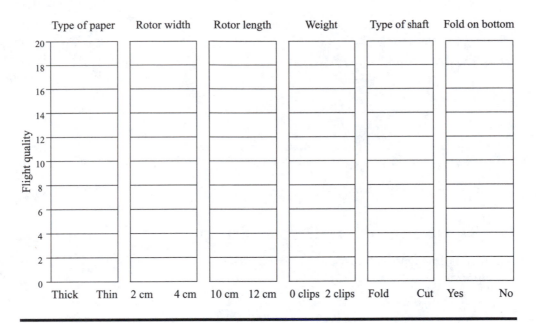

Figure 13.20 Plot of Helicopter Flight Quality

Q32. *Looking first at the time of flight, which factors have the greatest influence on flight time? Also note the factor level with the longest average flight time.*

Q33. *Repeat this with quality, remembering that the higher the value the better the quality.*

Q34. *Taking both time and quality into account, what levels of each factor would you choose in designing your helicopter? Does this correspond to one of the helicopters used in the experiment?*

Q35. *What should we do next to confirm our results and further improve the helicopter design?*

Q36. *What important concepts have we learnt from this experiment that we could apply to real problems?*

13.10   Data for the Helicopter Experiment

The results in Figure 13.21 were obtained from an experiment using the design in Figure 13.17.

Run	Time	Quality
1	2.53	1
2	1.15	3
3	2.54	5
4	1.59	2
5	1.34	1
6	1.99	4
7	1.67	3
8	1.99	2

Figure 13.21   Results for Helicopter Experiment

13.11   Exercises

1. A process is capable when it produces output which is
   a. all within the specification limits
   b. in statistical control
   c. out of statistical control, but all within the specification limits
   d. some outside the specification limits, but in statistical control
   e. all within the specification limits and in statistical control

2. In the helicopter experiment several factors were changed at a time because we wanted to
   a. test every combination of the six factors
   b. minimise the time doing the experiment
   c. take into account the interaction between factors
   d. minimise the variation from the experiment

3. Control charts are used to distinguish between
   a. common cause and special cause variation
   b. out of specification items and acceptable items

    c. good and bad items

    d. capable and incapable processes

4. The latest data value on the width of cardboard from the production process indicates a special cause above the upper control limit. Which of the following is the most appropriate action to take first?

    a. Look at all of today's data and break it down by hour, operator, etc. to see if any patterns can be found.

    b. Stop the process, and try to identify what was different about the current value.

    c. Turn the settings down to adjust for this high value.

    d. Run the process with different pressure settings to see the effects.

    e. Look at the raw paper materials to see if they have changed.

5. If a process is in the "brink of chaos" then to improve it we should first

    a. eliminate special causes

    b. experiment with the process settings to bring the process into control

    c. do nothing, since every component is in specification

    d. eliminate common causes

6. Process capability measures

    a. the percentage of process downtime

    b. the ability of the process to meet specifications

    c. whether quotas are being met

    d. the performance of the operators

7. When monitoring a process, the control limits need recalculating when

    a. a point is plotted which is outside the control limits

    b. new raw materials are used in the process

    c. a new operator is placed on the process

    d. the engineers have improved the process

    e. a special cause has been found

8. Decreasing the subgroup size when we collect samples for a control chart

    a. decreases the sensitivity of a control chart

    b. increases the sensitivity of a control chart

    c. has no effect unless you increase the frequency at which samples are collected

    d. has no effect at all

9. Which of the following best explains why specification limits are not drawn on control charts?

    a. operators have enough to worry about with control limits

    b. management is not interested in whether parts meet specifications

    c. control limits are more important than specification limits

    d. a control chart is used to spot special causes and these have nothing to do with specification limits

    e. control limits are the same as specification limits

10. Gertrude Williams is in charge of a printing press that prints customer information and designs onto cardboard boxes. Recently the machine has been producing blurred and running images that have been the reason for a large number of customer complaints over recent months. A control chart of the data shows no special causes.

a. What improvement strategies should Gertrude consider next?
b. Suppose Gertrude decides to carry out an experiment to investigate the cause of the problem. Engineers tell her that there are several factors that can influence printing clarity, namely dilution of the inks, ambient temperature, pressure setting of the rollers, speed of the press, type of cardboard, and the complexity of the design (number of colours, drawings, etc.). Discuss how you would set up an experiment to identify the important factors. You should clearly identify the factors and levels you would use and the way in which you would run the experiment. Write out a set of instructions so that a person could run the experiment in your absence.

11. Consider again Question 7 in Section 12.6. In an attempt to improve sales Neville is looking at the following options:
    1. Agree with Helen and suggest further training for the new salesperson. Also suggest that it would be a good idea to set up a sales training course for all of the sales staff, since it is quite a while since they all had a refresher course.
    2. Call the new salesperson in for a discussion and find out what he has been doing to sell the Z20A over the past 2 months, in the hope that you will be able to give him guidance on how to improve.
    3. Review the bonus scheme for the sales staff, placing more emphasis on payment by results.
    4. Break down the aggregated sales data for the Z20A by area, time of year, customer, and salesperson to see if you can see any patterns.
    5. Conclude that the Z20A is reaching the end of its product life cycle, and that it is time to develop a replacement model. Set up a market research survey to obtain new information on customer preferences, and report the results to Helen.
    Which option do you think Neville should choose? Give reasons for your choice and the rejection of the other options.

12. A team of line workers is looking at the thickness of a plastic laminate used to make plastic bags. Every hour they take four samples and measure the width (in hundredths of millimetres), using a laser as the plastic passes over a measuring bench. Figure 13.22 gives the measurements over 20 hours.
    a. Plot mean and range charts for these data.
    b. Calculate the control limits and draw them on the mean and range charts.
    c. What sort of strategies would you take to improve this process? List some ways in which you could apply these strategies to this process.

13. Jill Daley works in the despatch department at Pultron Rods in Hamilton. Part of her job is to ensure that orders are despatched to meet customer delivery dates. Customers normally order 5 days before they want the order. Pultron is always on the lookout to improve customer service, and Jill is setting up a control chart to help monitor delivery dates. She intends to collect data on 50 delivery times, recording the number of days ahead or behind the requested delivery date that customers receive their orders.
    a. Write down two reasons why this data might be difficult to obtain.
       Jill collects the 50 observations over 5 days and plots the data in Figure 13.23. The mean is 1.62 and the standard deviation 1.00.

Hour	Sample 1	Sample 2	Sample 3	Sample 4	Mean	Range
1	20.74	20.28	23.43	19.55	21.00	3.88
2	20.87	21.33	20.96	18.57	20.43	2.76
3	21.45	20.04	22.50	18.89	20.72	3.61
4	28.24	24.11	18.70	22.08	23.28	9.54
5	17.61	23.64	19.52	16.15	19.23	7.49
6	17.65	20.75	19.21	15.99	18.40	4.76
7	20.73	19.65	18.60	20.57	19.89	2.13
8	17.87	20.50	18.79	22.40	19.89	4.53
9	21.50	24.73	20.77	23.69	22.67	3.96
10	22.69	19.38	21.89	16.68	20.16	6.01
11	19.58	21.30	17.78	17.98	19.16	3.52
12	17.92	22.91	22.20	18.86	20.47	4.99
13	22.01	18.63	22.77	23.08	21.62	4.45
14	19.87	19.55	23.28	17.72	20.11	5.56
15	19.91	20.26	21.46	21.98	20.90	2.07
16	17.83	22.59	19.67	23.15	20.81	5.32
17	21.88	22.35	19.94	19.14	20.83	3.21
18	18.92	23.81	21.67	20.39	21.20	4.89
19	20.82	19.22	21.19	16.84	19.52	4.35
20	20.42	19.36	16.34	17.48	18.40	4.08
Totals					408.69	91.11

Figure 13.22 Thickness of Plastic Laminate

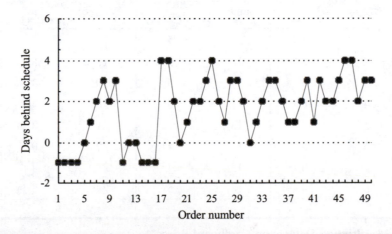

Figure 13.23 Run Chart of Delivery Times

b. Calculate the upper and lower control limits, and draw them on the chart.
c. What does the control chart tell us?
d. Which of the following options should Jill accept and which should she reject in her search for improvement? For each option, give a reason for your decision.
    i. Look at why the orders on the last day have all been late.
    ii. Collect more data and break down orders by month, day of week, time of day, type of order, etc. to determine any patterns.
    iii. Purchase a new production control system to help control production start dates and the ordering of component plastics.
    iv. Get upper management to send a memo to all production workers making them aware of the importance of keeping to planned production dates.
    v. Ring up customers and offer them discounts for placing orders more than 5 days ahead of delivery date.

14. Zia Sims is in charge of despatch, where she audits customer complaints. She has recently heard concerns about the printing quality of cardboard boxes supplied to customers, so she passed this information back to the printing department. Jo Arimu, the printing manager, decides to monitor the proportion of boxes with printing defects supplied to customers with a view to improving quality. Jo sets up a team from the department to look into the matter.
a. What should they do before they start collecting the data?
Later, they collect data on the number of defective boxes in 24 samples of 100 boxes taken from customers' orders. The data are given in Figure 13.24.
b. Calculate the upper and lower control limits, and plot them on a control chart.
c. What can you say about the process? What would you do next?
The printing team uses the control chart to monitor the process and the samples they take over the next few days have defective numbers of 5, 7, 4, and 12, respectively.
d. Which of the following strategies would you adopt first in your search for improvement, and why?

Sample	Number of defects	Sample	Number of defects	Sample	Number of defects
1	4	9	5	17	2
2	8	10	3	18	5
3	5	11	5	19	3
4	6	12	6	20	4
5	4	13	7	21	2
6	3	14	4	22	0
7	0	15	4	23	4
8	4	16	5	24	5

Figure 13.24   Printing Defects on Cardboard Boxes

i.   Stop the process immediately and see if you can identify any reason for the last high proportion of defective boxes. Look at the printing die, inks, operators, etc. to see if anything is different.

ii.   Have a word with the operator, tell him that he should adjust the printing machine as the ink is coming out too fast, and that he should be more alert in the future.

iii.   Look at all the defects over the period. Classify defects by colour of ink, size of print area, etc. and see if you can spot any patterns.

iv.   Order new ink from a different supplier. Try it out over a few days to see if it improves the printing quality.

v.   Since you have been considering buying a more modern printing press for a while now, bring the purchase forward and get it into operation as soon as possible.

15.   The manager of a large regional hospital is concerned with the waiting times experienced by patients arriving at accident and emergency (A&E). Hospital policy is that no patient should wait longer than 60 min before they see a doctor. The data in Figure 13.25 give the length of time in minutes (to the nearest 5 min) that 50 patients wait after arriving at A&E before they see a doctor.

20	40	35	25	20	100	40	30	50	55
130	60	50	15	10	45	85	70	40	110
240	70	90	20	30	65	45	50	10	15
110	20	70	45	35	30	45	55	15	85
50	45	65	70	40	45	40	55	65	45

Figure 13.25   Accident and Emergency Waiting Times

a.   Draw a stem and leaf diagram of the data and show the specification limits on your diagram.

b.   What percentage of data is beyond the specification limits?

c.   The mean waiting time of the sample is 54 min with a standard deviation of 38 min. If you assume the data are normally distributed, what proportion of patients would you expect to have to wait for more than 1 hour? Is this an appropriate estimate?

d.   What capability state is this process in?

Postscript

Decisions Based on Data

Throughout the book we have stressed the importance of basing decisions on data. You cannot understand variability unless you know the extent of the variability in the process you are studying. For this you need data. You cannot effectively improve a process, or solve a problem, unless you know the current state of the process or the scale of the problem. Again you need data. How can you understand the relationship between variables unless you look at the data? Useful forecasts of sales, unemployment, or commodity prices, for instance, cannot be made unless we know what has happened in the immediate past. Deciding whether a process is capable of meeting customer needs, or whether to apply common cause or special cause strategies, have to be based on an analysis of relevant data.

Too often such decisions are made on the basis of someone's hunch, or gut feeling. There is nothing wrong with having some hunch about a process or a problem, based on some previous experience perhaps, but it needs to be supported and backed by data. Otherwise there may be no basis for the decision, and this can prove to be very costly if at some later time the hunch turns out to be misguided, as is often the case.

Methods for collecting, presenting, and analysing data have been presented. Not all statistical topics of importance to managers, and future managers, can be covered in a book of this nature. Nor is it appropriate to do so. One of our aims has been to introduce you to methods that are, or should be, widely used in business. More importantly, our primary aim has been to get you thinking statistically about some of the issues faced in business. For some, more training may be needed. For instance, economists should further study regression (econometric methods) and forecasting. Those working in marketing need more on sampling methods, and on techniques that help you understand the factors underlying consumer preferences. Production managers will benefit from a greater understanding of statistical process control, and of the enormous benefits that can arise from using well-designed experiments.

Thinking Statistically: The Key Principles

Very often managers find it difficult to know how best to use data, or specify what data to collect. Often data are collected without knowing why, and much of it is never used. At other times data are used that has been collected for some other purpose, and are not really suitable for the problem or process under study. The following general principles will help you decide what data to collect and how to collect, present, and analyse it.

Define Clearly What Is Being Studied

If you are trying to improve a process, because too many errors are being made or there is too much variability or customers are complaining, then you should identify clearly what that process involves. Where does it begin and end? What are the stages involved in the process? Who or what is involved in the process? Simple top-down flowcharts are particularly useful in helping people understand a process, or processes (see Section 2.4). If good forecasts of sales are required for production planning purposes, be clear about what it is you have to forecast. Is it the sales of a particular product line, or of all products, or what? If you plan to survey your customers think carefully about what you want to find out. Planning what you are trying to achieve is a vital first step in any study or investigation.

Have Good Operational Definitions

Understand clearly and precisely what you want to observe and measure. This is not always easy, but essential if consistency in data is to be achieved. Decide how the data is going to be recorded. Use a well-designed, and tested, *check sheet*. Make sure the equipment used to collect data is adequate for the job. One company installing double-glazing sent one of their new sales people to measure the windows in a house. The person had not been told that he should measure the external, rather than the internal, dimensions of the window. When the new windows arrived, many of them did not fit. They had to be taken back to the factory and reworked. Essentially, it comes down to being adequately prepared to record or collect the required data, by knowing precisely what you want to observe or measure, and having the necessary equipment and facilities to do it.

Decide How to Collect the Data You Need

Usually it is not possible to collect information about every member of the population under study, or of all the items produced by some process. A sample of members or items is needed. In Chapter 6 the wood experiment was used to demonstrate the potential dangers of taking a judgment sample. No matter how conscientious or careful, it is almost impossible for an investigator to use judgment to choose a sample that is free of bias. Increasing the sample size simply makes things worse, in that the bias is likely to increase. Further, nothing can be said about the properties of the sample. With a random sample, however, we can determine these properties. For instance, we know that the distribution of the sample mean is approximately normal, is unbiased in the sense that the sample mean can be expected to equal the population mean in the long run, and that the standard error of the sample mean decreases as sample size increases. Thus, we can attach a margin of error to our estimate using the methods described in Chapter 7. Different ways of taking a random sample were briefly mentioned in Chapter 6. Although nonrandom sampling is not recommended, it is frequently used in practice because, in many situations, random samples are difficult or costly to obtain. For example, market research data often derives from nonrandom samples such as quota samples or self-selected samples. Data collected through a nonrandom sample is not necessarily worthless. It depends on the purpose of the study. Hotels, for instance, invariably leave a questionnaire in rooms asking for comments about the facilities and services provided. Information gained from such questionnaires can be useful in helping the hotel to better satisfy their customers' needs, even though the responses to such questionnaires do not necessarily constitute a random or representative sample of the hotel's customers.

Another important consideration in collecting data through a random sample is the question of nonresponse. A university carried out a survey of all its postgraduate students. Students were sent a questionnaire by e-mail, asked to fill it in, and to return it by e-mail. Only 28% of the students responded. What can we say about the 72% of students who did not respond? Were they the same as the group who did respond? Did they hold the same views? We do not know. The university may have been better advised to take a small random sample, stratified by factors such as gender and school of study, and follow up the initial e-mail questionnaire by further e-mails and personal interviews.

Another method of collecting data is through *experimentation*. Some of the principles involved were discussed in the context of the helicopter experiment in Chapter 13. Essentially an experiment involves some deliberate changes to the values, or levels, of factors that might have an important bearing on the outputs of the process. A careful choice of how these changes are made, and applying them in a random order, can give valuable insights into the probable effects of such changes. In an experiment the choice of which changes to make is under the control of the experimenter. In some situations, we are not free to make this choice. In an investigation of the relationship, if any, between smoking and lung cancer it would be unacceptable to choose which subjects smoked and which did not. All that can be done is to observe who smoked and whether they have lung cancer or not. This is an *observational study*, which can be useful for highlighting relationships and identifying possible causes. Only a properly designed and run experiment can demonstrate likely causation.

Collect Interval Data Whenever Possible

Greater information can be obtained from interval data than from attribute data. As explained in Chapter 5, the only property that attribute data possess is that of equivalence between items. It is not possible to say that one data value is so many units more than, or less than, another value. However, attribute data is usually easy and quick to collect. Avoid categorising interval data before analysing it. Knowing that a manufactured part is defective is not as informative as knowing the actual measurements of the part. Two defective parts can have very different characteristics, from only just unacceptable to totally unacceptable.

Think Carefully How to Present the Data

In the first week of each month a report on the costs of poor quality was presented to a General Manager. Nothing was ever done about these costs because the data were presented in such an unintelligible way that the General Manager failed to appreciate the importance of the figures. She was, after all, a busy person. It would have been better to bring attention to the main features of the data in a few well-presented tables and graphs; all the data could be placed at the end in an appendix, if need be. Advice on how to present data is given in Chapter 3, and two excellent books on this topic are given in the list of further reading at the end of the book (page 325). Reading these books will be well worthwhile.

Plot Your Data

The analysis of data should always begin by first plotting the data to see if there are any obvious patterns or relationships. With attribute data, use bar charts, pie charts, and Pareto diagrams. With interval data, a stem and leaf diagram will tell you about the location, spread, and shape of the data. A box plot is useful for comparing sets of interval data. These plots will also draw attention to unusual data points, or points that are different from the remainder of the data (outliers). Such points may be the result of mistakes and errors, which can usually be rectified, or they could be important genuine points. In this latter situation they would normally be analysed separately from the remainder of the data. An example is given by the debt collection data discussed in Section 1.3 and elsewhere. Here, there seems to be clear evidence that the January figures are different from the other months of the year. If the data have been collected in some time sequence then run charts should also be used, as patterns and trends over time, which are almost always important, are more easily detected. In studying relationships between variables, as in Chapters 8 and 9, scatterplots are essential. They indicate the type of regression model to fit, and provide a means of determining the appropriateness of the chosen model. Keep in mind that different graphs show different aspects of the data. Sometimes a run chart shows an important pattern in the data; in other instances a stem and leaf diagram, focusing on the shape of the data, will give a better understanding.

Keep it Simple

In collecting and plotting the data you may find that further analysis of the data is unnecessary. A Pareto diagram, for instance, will show where further work should, and should not, be concentrated. A check sheet, especially one with a visual element, may tell you where the problem lies without the need for any additional analysis. Some of the time, however, it will be necessary to carry out further analyses. It is important at this stage to keep in mind the purpose of your investigation. What you are trying to achieve in the analysis should have been identified at the outset. In a market research survey, are we really interested in whether there is a difference in the preferences between female and male customers? If not, then an analysis of any difference becomes unnecessary. If we are, then what decisions will be made if a real difference is found?

Take Advice when Necessary

Remember that this is an introductory text. There is much routine statistical analysis that you can do yourself, for example, designing a simple sample survey, calculating correlation coefficients and regression relationships, and using simple models to generate short-term forecasts from time series data. However, further study, and experience, will be necessary before you can confidently use many of the methods in more difficult situations—for instance, sampling an ill-defined population to take account of a range of factors within the population, carrying out a regression analysis on data that is nonlinear, forecasting future values for a series which contains complex patterns, or establishing control charts for a complicated process. All these require a level of expertise that you will not acquire from studying this text alone. Advice from a statistician will be necessary. Our aim has been to equip you with an understanding of statistical issues, which will allow you to make valuable contributions and suggestions, and to communicate effectively with the statistician.

What We Have Not Done

In many way this book is deceptively simple. It does not contain many formulae, and uses algebra only where necessary to represent a difficult concept succinctly. Do not be misled by this. Doing routine calculations and blindly using formulae that you do not really understand might look clever, but at best is rather pointless, and at worst potentially dangerous. You would not use a piece of complex machinery, or drive a car, without understanding what you are doing, and why you are doing it. Likewise, you should not use powerful statistical tools in the same way. It is important to know how to use data sensibly, and to know what not to do, either because it is inappropriate or because you do not have the necessary skills.

In Chapter 7 our emphasis was on estimating population parameters from sample data. In our opinion, most managers are, or should be, much more concerned with problems arising from the estimation of unknown population characteristics than in testing rather meaningless hypotheses. When trying to understand and explain variability we are, however, interested in examining whether differences we observe between factors in the process are meaningful, or significant, or whether such differences are due entirely to chance—in other words, in identifying factors that explain this variability. In Chapter 5 we used the chi-square test to examine whether apparent differences between attribute factors were real or not, for example, whether females had a greater preference for a product than males. In regression analysis we try to explain the variation in the response variable by one or more explanatory variables. In Chapters 8 and 9 we used the square of the (multiple) correlation coefficient for this purpose. More advanced techniques for examining whether an explanatory variable, or set of variables, is contributing significantly can be found in any book on regression analysis. When the explanatory variables are attribute variables we showed that dummy variables could be used in the regression analysis. However, there exist separate methods for analysing data where all, or almost all, these variables are attribute variables. These so-called *analysis of variance* methods are very powerful, but beyond the scope of this book. In the helicopter experiment, for example, we can use analysis of variance techniques to explain some of the variability in flying time in terms of the six factors in the experiment.

We do not think it necessary to explore the many aspects of formal probability theory or standard probability distributions that form the core of statistical theory. The one exception is the normal distribution. Due to its central role in statistical theory, it is important to fully understand the nature and purpose of this distribution. As to other distributions, however, we have made use of the key ideas from the t, chi-square, binomial, and Poisson distributions, and this is far more important than being able to perform largely meaningless probability calculations. This is the domain of the statistician rather than the manager.

These are some of the areas that we have chosen not to include, because we consider them unnecessary for the purpose of this book. There are, however, many other areas where valuable insights into complex data sets could be obtained from more advanced statistical analysis, but which are beyond the scope of this book. This is mainly in the fields of multivariate analysis, and more sophisticated analyses of time series. The problems of understanding complex relationships between large numbers of variables, and in being able to detect patterns and trends in data recorded over long periods of time, are the two main areas where the advice of a statistician is strongly recommended.

Statistical Computing

Computers are very good at doing the technical work. You do not have to know the mathematical formula for the standard deviation in order to calculate it. It is easily done on the computer, using software such as Excel. What is more important is that you understand what it is, and how to interpret and use it. Multiple regression models are easy to fit with appropriate computer software, even though the mathematics behind the technique is very complicated. Again it is sufficient to understand what multiple regression does, how to use it, the idea of fitting a model by least squares, and how to check the adequacy of the model and the assumptions made. We have used Excel in the book because it is simple to use, and our experience has shown that this (or similar spreadsheet packages) is what most managers use in business. For most of the material we have covered it is more than adequate. You can easily enter data into a spreadsheet, draw graphs and tables, and carry out many different analyses. Further, the outputs from Excel can be readily copied into a word processing package when writing a report, or into a graphics presentation package.

Specialist statistical packages are also available, which will do everything covered in this book and a lot more. One that is widely used for teaching at the level of this book is *Minitab*. Again data is entered into a spreadsheet, although calculations cannot be carried out in the spreadsheet, but in a separate window (or worksheet). It has functions for stem and leaf diagrams and box plots, for instance. Control charts are not difficult to do in Excel, but it is necessary to first calculate the limits and then add them to the run chart. Minitab can do this automatically. However, few managers in industry use Minitab routinely for their statistical work.

Probably the most widely used statistics package for the analysis of survey data is *SPSS* (*Statistics Package for the Social Sciences*). Like Minitab, *SPSS* also stores data in a rectangular array of rows and columns, much like a spreadsheet, but without the dynamic updating features of a normal spreadsheet. *SPSS* is ideal for the analysis of the very large datasets that arise from market research and other surveys. It contains a wide range of techniques for analysing complex multivariate models, and is available in an easy-to-use Windows version.

Appendix A

Introduction to Excel

A.1 Introduction

Anyone who has used a computer for more than just playing games will be aware of spreadsheets. A spreadsheet is a versatile computer program (package) that enables you to do a wide range of calculations *dynamically*, and create high quality graphs and charts. Microsoft Excel is the most widely used spreadsheet, and is available within the Microsoft Office suite of programs.

As you work through this introduction to Excel, it is a good idea to be at a computer so that you can try out various things as they are described. To get into Excel, simply double click on the Microsoft Excel icon if there is one on the computer desktop. Alternatively, click on the Start button in the bottom left corner of the screen, move the cursor to Programs to open up the programs menu, and then click on Microsoft Excel. Your computer should display the basic Excel screen shown in Figure A.1. You are now ready to use Excel.

The bulk of the screen is devoted to displaying a spreadsheet, which is one of many similar sheets that make up an Excel workbook. By default, the workbook is set up with three sheets, but this can be extended by creating additional sheets as required. In simple examples, you will often be able to organise your work on a single sheet, but for more complicated problems it may be more convenient to use several sheets. Different sheets are accessed by clicking on the appropriate Tab at the bottom of the screen. It is also possible to give the sheets more meaningful names, rather than the default names Sheet1, Sheet2, Sheet3, etc.

Across the top of the screen is a title bar that displays the name of the program followed by the name of the workbook ("Book1" is the default name, which will change when you save your data under some other, more meaningful name). On the right of the title bar are the usual buttons to minimise, maximise, and close a window. Below the title bar is the main menu bar that contains a series of pull-down menus, organised, as with most applications, into related groups of tasks. Selecting from the menu bar either opens up a submenu, or causes Excel to execute some task.

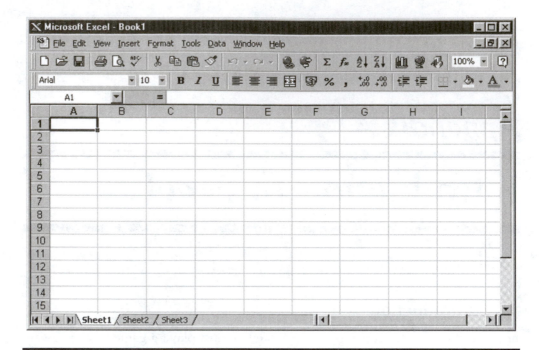

Figure A.1   Basic Excel Screen

Below the menu bar are two rows of toolbars that contain buttons (also called icons) for easy access to the more frequently used tasks. They allow you to execute the task by simply clicking on the appropriate button, rather than opening up a series of menus. Most of these buttons, such as *cut* or *paste*, are standard across many applications and work in exactly the same way as in, for example, Microsoft Word.

A.2   An Excel Spreadsheet

A spreadsheet works by laying out your data in a rectangular grid of rows and columns, in the same way that you would do if you were using a sheet of paper. For example, suppose you have collected quarterly data on home and export sales in the years 1998 and 1999. A natural way of displaying the data would be as shown in Figure A.2.

The data are laid out in seven columns and six rows, including row and column headings. As you can see in Figure A.1, in an Excel spreadsheet the rows are numbered from one upwards, and the columns are referred to by letters in alphabetical order. After the 26th column, they are labeled AA, AB, AC, ... then BA, BB, BC, ..., etc. When you have entered the data into Excel, what you see on the computer screen will look almost exactly like Figure A.2, apart from some blank columns after G, and blank rows below 6.

Any cell in the spreadsheet is identified by its column and row position. For example, the top left cell containing the word "Quarter" is cell A1, and the home sales in autumn 1999 (the value 24) is in cell E5. We often need to refer to a range of cells, which is a rectangular block of consecutive rows and columns.

	A	B	C	D	E	F	G
1	Quarter		1998			1999	
2		Home	Export	Total	Home	Export	Total
3	Spring	23	16	39	25	18	43
4	Summer	28	17	45	31	19	50
5	Autumn	21	11	32	24	12	36
6	Winter	20	13	33	21	15	36

Figure A.2   Quarterly Sales Data

For example, the 1998 data is contained in rows 3, 4, 5, and 6 of columns B, C, and D. In Excel, any range such as this is specified by the top left corner cell and the bottom right corner cell, separated by a colon. The 1998 data are therefore in the range B3:D6. Similarly, the 1999 export sales are in the range F3:F6, and all the sales data for 1999 are in the range E3:G6.

A.3   Entering and Editing Data

The cells of a spreadsheet can contain either numbers, as in cell B3, or characters (words or other symbols), as in cell A1. To enter data into the spreadsheet, click on the appropriate cell and type in the required contents and then press *Enter*. Immediately above the column headings of the spreadsheet, there is a formula bar that shows the active cell address, and the contents of that cell. In Figure A.1, the cursor is initially positioned on cell A1, and this cell address appears on the left of the formula bar. As you type an entry into any cell, the contents will appear after the = sign on the formula bar. In Figure A.1, cell A1 is initially empty; so, no contents are shown on the formula bar.

Try entering some data by setting up the spreadsheet exactly as shown in Figure A.2. As you enter data, you will see that when you press the *Enter* key the cursor automatically moves down to the cell below. You can change this automatic transition so that the cursor moves up, down, left or right (or not at all) after each data entry.

It is easy to change (edit) the contents of any cell. Click on the cell to be edited and then either:

- delete the entire contents of a cell by pressing the *Delete* key,
- type a new cell entry (there is no need to first delete the old entry as this will happen automatically), or
- modify the existing contents by positioning the cursor in the formula bar. Highlight the section to be changed, or click at the point where an insertion is to be made, and type in the new characters/data.

Practice these various editing operations by amending some of the cells in your spreadsheet. For example:

- click on cell A7 and enter the word "Tootal" (notice the cursor moves to cell A8 when you press *Enter*), and
- click on A7 again, highlight one of the o's in the formula bar and delete it (then press *Enter*).

To enter the same data repeatedly into a number of cells, we can use the *copy* and *paste* facility as follows.

- enter the data into the first cell(s)
- highlight these cells and press the *copy* button
- highlight the cells where the contents are to be copied, and press the *paste* button

Practice this by first deleting the contents of cells E2 to G2 (E2:G2), and then copy into these cells the contents of cells B2 to D2 (B2:D2).

A.4 Entering Formulae

If a spreadsheet were only able to hold input data, as on a sheet of paper, it would not be very useful. The power of a spreadsheet comes from the ability to enter formulae into the cells of the spreadsheet so that results can be calculated automatically. This allows the contents of any cell that contains a formula to be determined automatically from the contents of other cells. Furthermore, the calculation is dynamic in the sense that the result of the formula changes automatically whenever the values that the formula uses are changed.

For example, if the spreadsheet in Figure A.2 has appropriate formulae in the cells of columns D and G, the totals can be determined automatically from the sales data in cells B3:C6 and E3:F6. In particular, the total sales in the spring of 1998 is simply the sum of cells B3 and C3; so rather than enter the actual total (39) in cell D3, we can enter the formula

$$= B3 + C3$$

The = sign is necessary at the start of any formula so that Excel can distinguish a formula from a general text entry. After all, we might have wanted to put the characters "B3 + C3" into cell D3 as data.

All the other cells in column D contain similar formulae to that in cell D3, for example

(D4) = B4 + C4

(D5) = B5 + C5

(D6) = B6 + C6

When entering these formulae, the cell reference in brackets on the left is not required, as the cursor will be positioned on the cell concerned.

Input these formulae into the appropriate cells. Notice that the formula appears on the formula bar as you type it in. Notice also that you do not see the formula in the cell itself, only the result of the formula. The spreadsheet therefore appears no different from how it did before, but the contents of column D are now dynamically linked to those of columns B and C.

It can be time-consuming to type in every formula individually, but the process can be made much simpler. When essentially the same formula is being put in each cell, it can be copied from one cell to another using *copy* and *paste*, just as data can be copied. The formula that we have just entered into column D is simply the sum of the two cells immediately to the left. If the formula in cell D3 were therefore copied and pasted to cell D4, the original reference to cells B3 and C3 will be automatically updated to refer to cells B4 and C4. Copying a formula across or down the spreadsheet will automatically update the row and column references within the formula. So, for example, if the formula in cell D3 is copied to cell G3, the formula will become

$$= E3 + F3$$

Likewise, if copied to cell G7, it will become

$$= E7 + F7$$

As the formulae that we require in column G are essentially the same as in column D (i.e., the sum of the two cells immediately to the left), any one of the formulae in column D can be copied and pasted to cells G3:G6 in one operation.

Try this for yourself.

> Click on (say) cell D3 and press the *Copy* button
> Highlight cells G3:G6 and press the *Paste* button

Finally, decide what formula is required to give the column total in cell B7 and enter this formula into cell B7. As essentially the same formula will be used in each of cells B7:H7, the formula in B7 can copied and pasted to cells C7:G7.

A.5 Absolute Cell References

When copying a formula, we usually want the cell references within the formula to be automatically updated for different rows and/or columns. However, there are occasions when we do not want this to happen. For example, we might want a reference to cell D7 to remain as that no matter where the formula occurs. We can prevent cell references from being updated by inserting a $ symbol before either a row number or a column letter within a cell reference. For example, if a cell is referred to as $D7, then copying a formula containing this cell reference will *fix* the column at D, which will not be changed when the formula is copied to different columns. The row reference, however, will be updated as the formula is copied to different rows. Similarly, referring to a cell as D$7 will fix the row reference (i.e., it will always be copied as row 7), but allow the column to be updated. Fixing both row and column references, e.g., $D$7, causes the formula to refer to cell D7, no matter where it is copied.

A quick way of fixing a row or a column or both is to type in the required cell reference (without the dollar signs) and use the F4 key repeatedly to "toggle" through the different $ combinations. For example, having typed D7, pressing the F4 key once changes this to $D$7; successive presses of the F4 key produce D$7, $D7, D7, etc.

To see an example of this, suppose we wish to work out what percentage of each year's sales were home and export; for example the *Home* percentage in 1998 would be 92 × 100/149 = 61.7%. Suppose we wish to calculate these percentages in row 8. To avoid having to type the same formula repeatedly, we can enter it once in cell B8 and then copy to cells C8:D8. The formula we use in cell B8 would be

= B7*100/$D7   [= B7*100/$D$7 would produce the same effect]

Can we copy this same formula into cells E8:G8 to give the 1999 percentages? If not, what formula do we need in cell E8, which can then be copied to cells F8:G8?

Enter the correct formulae into cells B8 and E8, then copy and paste to complete row 8 in Figure A.3.

	A	B	C	D	E	F	G
	⋮	⋮	⋮	⋮	⋮	⋮	⋮
7	Total	92	57	149	101	64	165
8	Percent	61.74497	38.25503	100	61.21212	38.78788	100

Figure A.3   Calculation of Percent Row

A.6   Inserting Rows and Columns and Formatting Cells

As a final exercise, suppose we wish to calculate the quarterly percentage breakdown in total annual sales for each year. We could perform these calculations to the right of the data, in columns H and I. However, it would be better to put the 1998 percentages beside the 1998 data (i.e., between columns D and E) and the 1999 percentages after column G. We need, therefore, to insert a new blank column after column D. To do this first position the cursor in the column immediately to the right of where the new column will be, and then click on *Insert* on the top menu bar, and choose *Columns*. A blank column will appear to the left of the cursor, and the columns to the right will be relabeled. Rows can be inserted by a similar procedure.

Insert the blank column, which will now be column E, and calculate the percentage quarterly breakdown by entering and copying the appropriate formula. Notice that the formulae in both columns are essentially the same so if the correct formula is entered in cell E3, it can be copied to E4:E7 and I3:I7. Finally, type in the heading "Percent" in cell A8, and copy it to cells E2 and I2.

You will see that the percentage figures that you have calculated are given (by default) to 5 decimal places. In reality, this is unnecessary accuracy, and probably 1 decimal place is more than adequate. The contents of any cell can be displayed in a wide variety of ways to suit the purpose of the spreadsheet. The format of cells is adjusted by selecting *Format* from the top menu bar, and choosing *Cells*. This allows you to change features such as alignment (left, right, or centred), font, and decimal places.

More directly, you can set the alignment within any cell using the usual alignment buttons ≡ ≡ ≡ and increase/decrease the number of decimal places by clicking on the following buttons. ⁺₀₈ ·₀₈

If you centre all the cells (apart from column A), and reduce the percentage figures to 1 decimal place, your spreadsheet should now look like that in Figure A.4.

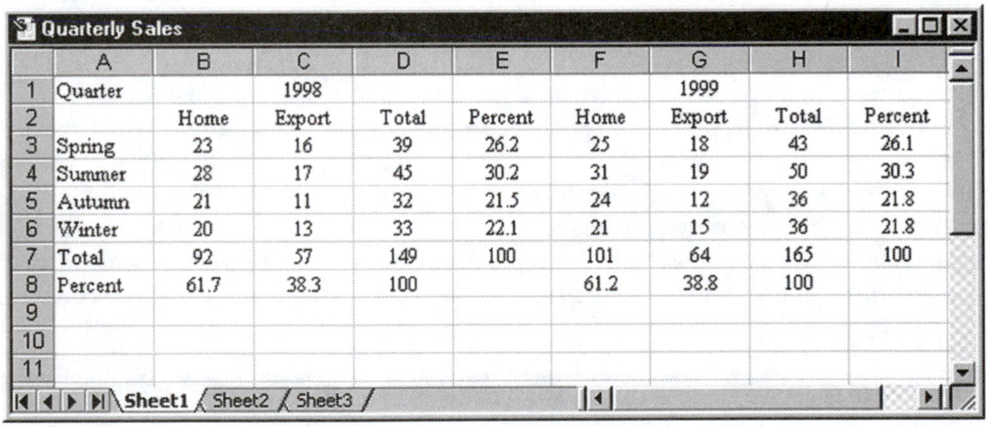

	A	B	C	D	E	F	G	H	I
Quarterly Sales									
1	Quarter		1998				1999		
2		Home	Export	Total	Percent	Home	Export	Total	Percent
3	Spring	23	16	39	26.2	25	18	43	26.1
4	Summer	28	17	45	30.2	31	19	50	30.3
5	Autumn	21	11	32	21.5	24	12	36	21.8
6	Winter	20	13	33	22.1	21	15	36	21.8
7	Total	92	57	149	100	101	64	165	100
8	Percent	61.7	38.3	100		61.2	38.8	100	
9									
10									
11									

Sheet1 / Sheet2 / Sheet3 /

Figure A.4   Final Spreadsheet

A.7   Standard Functions

Suppose we need to enter a formula that calculates the total of 10 values entered in cells A1:A10. It would be very tedious to have to type in the following formula:

$$= A1 + A2 + A3 + A4 + A5 + A6 + A7 + A8 + A9 + A10$$

Fortunately, there is a shortcut. We can use a standard function for the sum of a range of cells. In the above case we could abbreviate the formula to

$$= SUM(A1:A10)$$

The cells we wish to sum do not have to be in a single row or column; they could cover a number of adjacent rows and columns. For example, the total 1998 sales in cell D7 of Figure A.2 could be written as

$$= \text{SUM(B3:C6)}$$

Similarly, a number of distinct ranges could be included within the SUM function. For example, a formula for the overall sales for the two years could be written as

$$= \text{SUM(B3:C6, E3:F6)}$$

SUM is just 1 of over 200 standard functions, the majority of which you will probably have no need to use. The more commonly used functions are

SUM	Sum of a range of cells
MAX	Maximum of a range of cells
MIN	Minimum of a range of cells
SQRT	Square root of a cell or value
AVERAGE	Average (arithmetic mean) of a range of cells
STDEV	Standard deviation of a range of cells
MEDIAN	Median of a range of cells

Other, more specialised, functions will be introduced as they are required.

All functions that you will use require one or more *arguments* in brackets following the function name. These arguments almost always specify the cell or range of cells to which the calculation applies. If you know what arguments a particular function requires, you can simply type in the function name and its arguments as we did with the SUM function above. A simpler way, however, is to use the *Insert* menu and select *Function* as shown in Figure A.5. This brings up

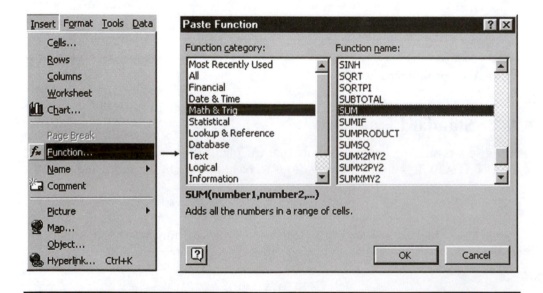

Figure A.5   Selecting a Function

the *Paste Function* dialogue window from which you select the required function from the relevant category. This then opens a dialogue box for that function which prompts you for the required arguments. The dialogue box for the *SUM* function is shown in Figure A.6. An alternative to selecting function from the *Insert* menu is to click the function button f_x.

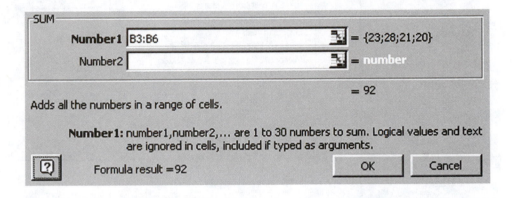

Figure A.6   *SUM* **Function Dialogue Box**

A.8   Excel Add-Ins

Most of the basic statistical calculations, such as an average or standard deviation, are available as functions in Excel. However, advanced statistical analyses that involve more complex calculations are provided in a *Data Analysis* tool-pack that is available in the *Tools* section of the menu. However, along with other more advanced features of Excel, *Data Analysis* is not present in the basic *Tools* menu, and must be installed if required. These are the so-called Excel *Add-Ins*. Once a particular Add-In has been installed, it remains in the *Tools* menu until it is removed; so, it only needs to be installed once.

To install *Data Analysis* (or any other Add-In) open the *Tools* menu and select *Add-Ins*. From the list of available Add-Ins, check the *Analysis ToolPak* box and click on OK as shown in Figure A.7.

When you have added in the *Analysis ToolPak, Data Analysis* will then appear in the *Tools* menu. Clicking on *Data Analysis* will open the window shown in Figure A.8. You can then select any of the available analysis tools, and pressing OK will open up a further dialogue box specific to the analysis required.

A.9   Graphs and Charts

Excel can be used to create a wide range of high quality graphs and charts from data contained in a spreadsheet. The simplest way to do this is to use the *Chart Wizard*, which takes you step by step through the process of creating a chart of the type that you want. The best way to fully appreciate the many graph and chart drawing facilities within Excel is to experiment with a simple set of data. To get you started, however, we will go through the various stages that you follow when using the chart wizard.

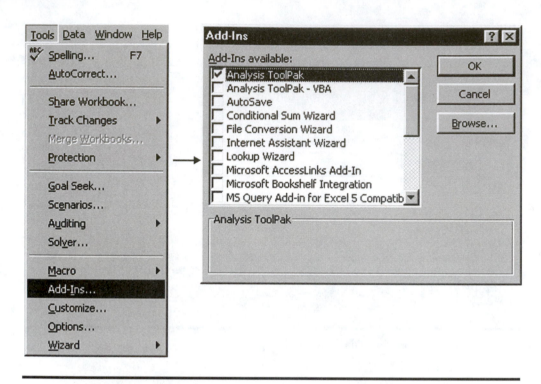

Figure A.7   Installing Data Analysis

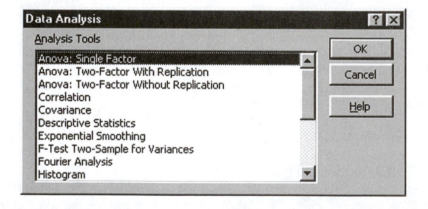

Figure A.8   Data Analysis Selection Window

First, select the *Chart Wizard* by clicking the following button. This will take you through a four-step process to build a chart of the form you want. We will illustrate below each of the four steps, along with the relevant dialogue box.

From the list on the left in Figure A.9, choose one of the standard chart types. As you click on a particular type the various options within that chart type are displayed in pictorial form on the right. When you have decided on a chart type and a particular form of that chart, click on the relevant picture to the right.

Suppose you want a vertical (i.e., column) bar chart. There are seven options of that type. The three styles on the top row have clustered, stacked, and percentage bars with no depth. On the second row are the same three forms but

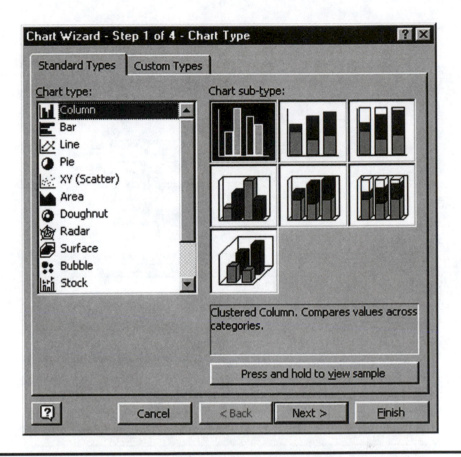

Figure A.9   Step 1: Specify the Chart Type

with three-dimensional bars. On the third row is a chart with the bars in three-dimensional space. As you click on any picture, a brief description is given, and if you have highlighted some data before selecting the chart wizard, you can get a preview of how your chart will look by pressing and holding the *View Sample* button. Having chosen the graph type, click the *Next* button.

If you have not highlighted your data before entering chart wizard, you now specify the *Data range* in Figure A.10 by highlighting it on the spreadsheet. When you do this a picture of your data will appear in the area at the top. Try this yourself by highlighting the range A2:C6 of the sales spreadsheet in Figure A.4.

We have included the titles (row 2 and column A) in the data range; so these appear on the graph as labels on the *x*-axis and the graph legend. See what happens to the graph if you click the *Rows* radio button to specify that the data are organised by rows rather than columns. In practice, the choice of data in rows or columns will depend on which feature of the data you wish to emphasise in the graph. Having specified the source data, click the *Next* button, or go *Back* to the first step to change the graph type.

In step 3, shown in Figure A.11, you modify the display features of the graph, and add any titles. The dialogue window contains a number of tabs that you choose to set the various graph characteristics. For example, the graph titles are incorporated by clicking on each title box in turn and typing in the required text.

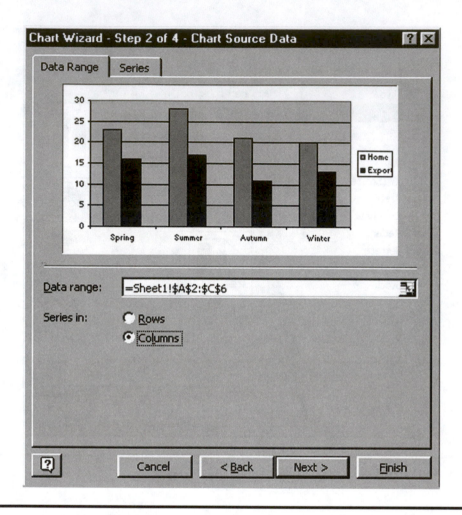

Figure A.10 Step 2: Specify the Source Data

After a short delay, the titles appear on the graph itself as shown. By default, the graph is drawn with horizontal (Y-axis) gridlines. Using the gridlines tab, these can be removed and/or vertical (X-axis) gridlines added. The legend tab allows the legend box (specifying the meaning of the various colours used for the bars) to be included or not, and if included, where it is positioned. Clicking on *Next* takes you to the final step, or *Back* to step 2.

The final dialogue box, shown in Figure A.12, simply controls the location of the graph. It can be included in the workbook as a complete sheet in its own right, or as an object within a specified sheet. Check the appropriate radio button and if it is to be a new sheet, enter an appropriate name by replacing the default name *Chart1*.

A.10 Printing and Saving a Workbook

Once a spreadsheet has been created, including any graphs, it can be printed and/or saved using the appropriate selections from the *File* menu. When saving, the entire workbook will be saved.

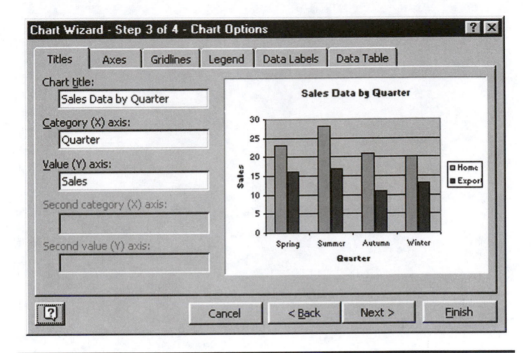

Figure A.11   Step 3: Specify the Chart Options

Figure A.12   Step 4: Specify the Chart Location

When printing, you may print a selection of a sheet (i.e., a highlighted range), a chart, a whole sheet, or an entire workbook. When printing a chart, first click on the chart in question before entering the print menu. You will then just print the selected chart rather than the spreadsheet itself.

Appendix B
Statistical Tables

B.1 Normal Distribution

The table gives the probability (*P*) that an item is greater than a value *z* for a standard normal distribution.

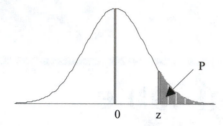

z	0.00	0.01	0.02	0.03	0.04	0.05	0.06	0.07	0.08	0.09
0.0	0.5000	0.4960	0.4920	0.4880	0.4840	0.4801	0.4761	0.4721	0.4681	0.4641
0.1	0.4602	0.4562	0.4522	0.4483	0.4443	0.4404	0.4364	0.4325	0.4286	0.4247
0.2	0.4207	0.4168	0.4129	0.4090	0.4052	0.4013	0.3974	0.3936	0.3897	0.3859
0.3	0.3821	0.3783	0.3745	0.3707	0.3669	0.3632	0.3594	0.3557	0.3520	0.3483
0.4	0.3446	0.3409	0.3372	0.3336	0.3300	0.3264	0.3228	0.3192	0.3156	0.3121
0.5	0.3085	0.3050	0.3015	0.2981	0.2946	0.2912	0.2877	0.2843	0.2810	0.2776
0.6	0.2743	0.2709	0.2676	0.2643	0.2611	0.2578	0.2546	0.2514	0.2483	0.2451
0.7	0.2420	0.2389	0.2358	0.2327	0.2296	0.2266	0.2236	0.2206	0.2177	0.2148
0.8	0.2119	0.2090	0.2061	0.2033	0.2005	0.1977	0.1949	0.1922	0.1894	0.1867
0.9	0.1841	0.1814	0.1788	0.1762	0.1736	0.1711	0.1685	0.1660	0.1635	0.1611
1.0	0.1587	0.1562	0.1539	0.1515	0.1492	0.1469	0.1446	0.1423	0.1401	0.1379
1.1	0.1357	0.1335	0.1314	0.1292	0.1271	0.1251	0.1230	0.1210	0.1190	0.1170
1.2	0.1151	0.1131	0.1112	0.1093	0.1075	0.1056	0.1038	0.1020	0.1003	0.0985
1.3	0.0968	0.0951	0.0934	0.0918	0.0901	0.0885	0.0869	0.0853	0.0838	0.0823
1.4	0.0808	0.0793	0.0778	0.0764	0.0749	0.0735	0.0721	0.0708	0.0694	0.0681
1.5	0.0668	0.0655	0.0643	0.0630	0.0618	0.0606	0.0594	0.0582	0.0571	0.0559
1.6	0.0548	0.0537	0.0526	0.0516	0.0505	0.0495	0.0485	0.0475	0.0465	0.0455
1.7	0.0446	0.0436	0.0427	0.0418	0.0409	0.0401	0.0392	0.0384	0.0375	0.0367
1.8	0.0359	0.0351	0.0344	0.0336	0.0329	0.0322	0.0314	0.0307	0.0301	0.0294
1.9	0.0287	0.0281	0.0274	0.0268	0.0262	0.0256	0.0250	0.0244	0.0239	0.0233
2.0	0.0228	0.0222	0.0217	0.0212	0.0207	0.0202	0.0197	0.0192	0.0188	0.0183
2.1	0.0179	0.0174	0.0170	0.0166	0.0162	0.0158	0.0154	0.0150	0.0146	0.0143
2.2	0.0139	0.0136	0.0132	0.0129	0.0125	0.0122	0.0119	0.0116	0.0113	0.0110
2.3	0.0107	0.0104	0.0102	0.0099	0.0096	0.0094	0.0091	0.0089	0.0087	0.0084
2.4	0.0082	0.0080	0.0078	0.0075	0.0073	0.0071	0.0069	0.0068	0.0066	0.0064
2.5	0.0062	0.0060	0.0059	0.0057	0.0055	0.0054	0.0052	0.0051	0.0049	0.0048
2.6	0.0047	0.0045	0.0044	0.0043	0.0041	0.0040	0.0039	0.0038	0.0037	0.0036
2.7	0.0035	0.0034	0.0033	0.0032	0.0031	0.0030	0.0029	0.0028	0.0027	0.0026
2.8	0.0026	0.0025	0.0024	0.0023	0.0023	0.0022	0.0021	0.0021	0.0020	0.0019
2.9	0.0019	0.0018	0.0018	0.0017	0.0016	0.0016	0.0015	0.0015	0.0014	0.0014
3.0	0.0013	0.0013	0.0013	0.0012	0.0012	0.0011	0.0011	0.0011	0.0010	0.0010

B.2 Percentage Points of the *t*-Distribution

The table gives the 90%, 95%, 98%, and 99% points of the *t*-distribution for different numbers of degrees of freedom (df).

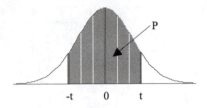

df	P = 90%	P = 95%	P = 98%	P = 99%	df	P = 90%	P = 95%	P = 98%	P = 99%
1	6.314	12.706	31.821	63.656	32	1.694	2.037	2.449	2.738
2	2.920	4.303	6.965	9.925	33	1.692	2.035	2.445	2.733
3	2.353	3.182	4.541	5.841	34	1.691	2.032	2.441	2.728
4	2.132	2.776	3.747	4.604	35	1.690	2.030	2.438	2.724
5	2.015	2.571	3.365	4.032	36	1.688	2.028	2.434	2.719
6	1.943	2.447	3.143	3.707	37	1.687	2.026	2.431	2.715
7	1.895	2.365	2.998	3.499	38	1.686	2.024	2.429	2.712
8	1.860	2.306	2.896	3.355	39	1.685	2.023	2.426	2.708
9	1.833	2.262	2.821	3.250	40	1.684	2.021	2.423	2.704
10	1.812	2.228	2.764	3.169	41	1.683	2.020	2.421	2.701
11	1.796	2.201	2.718	3.106	42	1.682	2.018	2.418	2.698
12	1.782	2.179	2.681	3.055	43	1.681	2.017	2.416	2.695
13	1.771	2.160	2.650	3.012	44	1.680	2.015	2.414	2.692
14	1.761	2.145	2.624	2.977	45	1.679	2.014	2.412	2.690
15	1.753	2.131	2.602	2.947	46	1.679	2.013	2.410	2.687
16	1.746	2.120	2.583	2.921	47	1.678	2.012	2.408	2.685
17	1.740	2.110	2.567	2.898	48	1.677	2.011	2.407	2.682
18	1.734	2.101	2.552	2.878	49	1.677	2.010	2.405	2.680
19	1.729	2.093	2.539	2.861	50	1.676	2.009	2.403	2.678
20	1.725	2.086	2.528	2.845	55	1.673	2.004	2.396	2.668
21	1.721	2.080	2.518	2.831	60	1.671	2.000	2.390	2.660
22	1.717	2.074	2.508	2.819	65	1.669	1.997	2.385	2.654
23	1.714	2.069	2.500	2.807	70	1.667	1.994	2.381	2.648
24	1.711	2.064	2.492	2.797	75	1.665	1.992	2.377	2.643
25	1.708	2.060	2.485	2.787	80	1.664	1.990	2.374	2.639
26	1.706	2.056	2.479	2.779	85	1.663	1.988	2.371	2.635
27	1.703	2.052	2.473	2.771	90	1.662	1.987	2.368	2.632
28	1.701	2.048	2.467	2.763	95	1.661	1.985	2.366	2.629
29	1.699	2.045	2.462	2.756	100	1.660	1.984	2.364	2.626
30	1.697	2.042	2.457	2.750	120	1.658	1.980	2.358	2.617
31	1.696	2.040	2.453	2.744	∞	1.645	1.960	2.326	2.576

B.3 Percentage Points of the χ^2-Distribution

The table gives the 10%, 5%, 2%, and 1% points of the χ^2-distribution for different numbers of degrees of freedom (df).

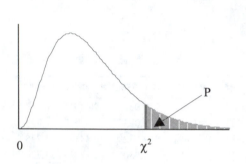

df	P = 10%	P = 5%	P = 2%	P = 1%
1	2.71	3.84	5.41	6.63
2	4.61	5.99	7.82	9.21
3	6.25	7.81	9.84	11.34
4	7.78	9.49	11.67	13.28
5	9.24	11.07	13.39	15.09
6	10.64	12.59	15.03	16.81
7	12.02	14.07	16.62	18.48
8	13.36	15.51	18.17	20.09
9	14.68	16.92	19.68	21.67
10	15.99	18.31	21.16	23.21
11	17.28	19.68	22.62	24.73
12	18.55	21.03	24.05	26.22
13	19.81	22.36	25.47	27.69
14	21.06	23.68	26.87	29.14
15	22.31	25.00	28.26	30.58
16	23.54	26.30	29.63	32.00
17	24.77	27.59	31.00	33.41
18	25.99	28.87	32.35	34.81
19	27.20	30.14	33.69	36.19
20	28.41	31.41	35.02	37.57
21	29.62	32.67	36.34	38.93
22	30.81	33.92	37.66	40.29
23	32.01	35.17	38.97	41.64
24	33.20	36.42	40.27	42.98
25	34.38	37.65	41.57	44.31
26	35.56	38.89	42.86	45.64
27	36.74	40.11	44.14	46.96
28	37.92	41.34	45.42	48.28
29	39.09	42.56	46.69	49.59
30	40.26	43.77	47.96	50.89
40	51.81	55.76	60.44	63.69
50	63.17	67.50	72.61	76.15
60	74.40	79.08	84.58	88.38
70	85.53	90.53	96.39	100.43
80	96.58	101.88	108.07	112.33
90	107.57	113.15	119.65	124.12
100	118.50	124.34	131.14	135.81

B.4 Random Numbers

70	71	42	34	1	74	88	50	53	94	22	16
52	5	30	51	18	11	4	67	22	80	13	32
69	2	3	77	29	67	40	41	100	13	92	19
98	78	82	48	43	52	19	75	98	75	88	23
73	3	53	81	72	95	65	71	1	69	2	21
86	19	20	95	98	53	32	87	79	1	35	12
47	29	13	71	23	41	17	88	9	59	52	61
76	16	8	67	66	82	26	34	33	91	88	31
72	9	60	10	90	94	35	80	26	70	69	98
13	4	4	40	81	89	34	69	39	56	36	47
5	57	88	29	89	28	66	44	94	48	38	3
43	97	3	7	87	11	14	20	71	39	62	31
59	52	29	4	92	14	85	66	50	44	57	44
9	40	24	18	24	20	82	58	66	48	76	23
40	66	45	90	2	36	16	96	52	58	4	65
92	99	65	71	67	43	21	51	96	59	44	6
86	30	89	97	15	55	24	70	43	56	43	87
46	48	40	74	9	70	39	48	85	58	85	41
79	85	9	38	96	11	25	89	43	75	6	27
52	57	35	28	71	63	68	67	57	58	60	48
4	69	6	36	89	8	22	58	10	80	83	61
90	83	78	47	26	52	2	9	36	20	42	87
24	76	52	91	92	60	35	9	84	22	46	33
22	56	83	19	69	40	91	60	74	31	15	91
12	17	30	86	49	70	75	94	59	89	73	85
21	75	2	45	32	69	49	65	68	71	94	62
47	84	67	37	37	41	64	55	64	33	47	74
74	56	36	86	29	76	3	80	39	7	28	45
85	54	82	33	42	32	65	82	19	44	18	18
93	7	68	14	54	51	47	44	11	47	13	2
3	99	36	18	25	99	42	23	84	98	85	84

B.5   Constants for Control Charts

Subgroup size	A_2	D_3	D_4
2	1.880	0	3.267
3	1.023	0	2.574
4	0.729	0	2.282
5	0.577	0	2.114
6	0.483	0	2.004
7	0.419	0.076	1.924
8	0.373	0.136	1.864
9	0.337	0.184	1.816
10	0.308	0.223	1.777

Bibliography and Further Reading

Three books that we have referred to in the book, and which we strongly recommend to you as giving the wider picture, are

Fourth Generation Management, by Brian L. Joiner, New York: McGraw-Hill, 1994.
The Team Handbook, by Peter R. Scholtes (and others), Madison: Oriel Consulting Inc, 1988.
The Leader's Handbook, by Peter R. Scholtes, New York: McGraw-Hill, 1998.

Many books have been written about the management methods of Dr. W. Edwards Deming, and are well worth studying. As a start, we suggest:

Out of the Crisis, by W. Edwards Deming, Cambridge: MIT Center for Advanced Engineering Study, 1986.
The Deming Management Method, by Mary Walton, New York: Putnam Publishing, 1986.

Short introductory statistics texts, focused on quality improvement, which will supplement the material in Chapters 2, 3, and 11 to 13 in particular, are

Guide to Quality Control, by K. Ishikawa, Tokyo: Asian Productivity Organization, 1982.
Statistical Methods for Quality Improvement, by H. Kume, Tokyo: AOTS Chosakai, 1985.

Two books recommended in Chapter 3 for their advice and wisdom on how to effectively present data are

A Primer in Data Reduction, by A.S.C. Ehrenberg, New York: John Wiley & Sons, 1982.
The Visual Display of Quantitative Information, by E.R. Tufte, Cheshire: Graphics Press, 1983.

There are many books written on the analysis of attribute data, usually under the guise of categorical data analysis or contingency table analysis. A good text to start with is

An Introduction to Categorical Data Analysis, by A. Agresti, New York: John Wiley & Sons, 1996.

For those interested particularly in market research and requiring a fuller discussion of the principles and methods of survey sampling, we recommend:

Sample Survey Principles and Methods, by V. Barnett, London: Edward Arnold, 1991.

The classic text on regression analysis, which was first published in 1966 and is now in its third edition, is

Applied Regression Analysis, by N.R. Draper and H. Smith, New York: John Wiley & Sons, 1998.

There are a number of good books on forecasting aimed at business practitioners, but often running to many hundreds of pages. For further study we would recommend:

Time Series Models for Business and Economic Forecasting, by P.H. Franses, London: Cambridge University Press, 1988.
Sales Forecasting Management: Understanding the Techniques, Systems, and Management of the Sales Forecasting Process, by J.T. Mentzer and C.C. Bienstock, London: Sage Publications, 1998.
Forecasting Methods and Applications, by S. Makridakis, S.C. Wheelwright, and R.J. Hyndman, New York: John Wiley & Sons, 1998.

A more statistical introduction to time series is given in:

The Analysis of Time Series: An Introduction, by C. Chatfield, London: Chapman and Hall, 1996.

A short, but excellent, introduction to statistical process control and its role in helping us understand variation is

Understanding Variation: The Key to Managing Chaos, by D.J. Wheeler, Knoxville: SPC Press, 1993.

For a more detailed treatment of statistical process control we recommend:

Understanding Statistical Process Control, by D.J. Wheeler and D.S. Chambers, Knoxville: SPC Press, 1992.

The helicopter exercise in Section 13.9 gives a brief introduction to industrial experimentation. The following book provides an excellent introduction to the many issues and concepts involved in experimentation, as well as providing details on how to design and analyse experiments.

Quality Improvement through Planned Experimentation, by R.D. Moen, T.W. Nolan, and L.P. Provost, New York: McGraw-Hill, 1999.

At a more advanced level, we recommend the classic text:

Statistics for Experimenters, by G.E.P. Box, W.G. Hunter, and J.S. Hunter, New York: John Wiley & Sons, 1978.

Information on the statistical software packages *Minitab* and *SPSS* can be found on their Internet sites.

For *Minitab*   http://www.minitab.com/
For *SPSS*   http://www.spss.com/

Index